Invitation to STATISTICS

Invitation to STATISTICS

Gavin Kennedy

MARTIN ROBERTSON · OXFORD

First published in 1983 by Martin Robertson & Company Ltd.,
108 Cowley Road, Oxford OX4 1JF

Kennedy, Gavin
Invitation to statistics.
1. Mathematical statistics
I. Title
519.5 QA276

ISBN 0-85520-553-9
ISBN 0-85520-554-7 Pbk

Typeset by Communitype, Oadby, Leicester
Printed and bound in Great Britain
by T.J. Press Ltd, Padstow

For Kim

Contents

Preface

The reader is invited to join the author in a relatively painless tour of statistics and statistical methods. Statistics – for many students and general readers alike – is perplexing. However, it can be an interesting experience if you can get beyond the seemingly endless number crunching, which too often demoralizes beginners, to an appreciation of its major scientific purposes. This books attempts to do just that. It is not a conventional statistics primer in elementary manipulations nor is it another text on 'uses and abuses' themes. Rather it is about the place of statistics in the wider scientific approach to human welfare. It is primarily aimed at 'beginners' (particulary those without mathematical skills – though having them is not a disqualification!) who want to get an overview of the subject before they embark on a statistics course for a social science or professional qualification. In other words, the book is intended for those who are likely to become perplexed by statistics. It should also be of interest to the general reader who finds adages about 'lies and statistics' inadequate to cope with the flows of information and assertion common in our multi-media society. In addition, 'old hands' might also enjoy perusing its contents.

I have deliberately refrained from demonstrating statistical manipulations in the text (though inevitably some have crept in): *this is not a 'how to do it' book; it is more about 'why it is done'*. I aim to induce a response akin to 'ah-aah' ('so that's what it means') rather than promote competence in statistical techniques. This is not to relieve the reader of the difficult bits. Far from it; in my view it is easier to calculate many (if not most) of the standard statistical manipulations than it is to grasp the concepts implicit in the statistical method. This is even more true with the advent of computer-based statistics packages that 'save' the student from arid chains of number crunching, and merely require an ability to press buttons in the right sequence. 'Saving' the student from boring work can sometimes induce an apathy about what it all means. Hence anyone reading this book to 'escape' from intellectual effort is going to be disappointed, though effort expended here to make the subject more meaningful is likely to reduce the effort required later to learn about statistical techniques. Not all authorities will agree with the approach I have adopted – perhaps out of a genuine belief that in the long run it does (all?) students good if they are dropped in at the deep end. In my defence, I believe that my approach is more likely to produce an appreciation of the statistical method.

I am firmly convinced that statistical competence is an essential component of a tertiary education and is absolutely necessary for making informed choices, for assessing the conduct and views of others, and for making workable assumptions about the wider world we live in. Much human discourse that threatens to break out into conflict or confrontation might be avoided, or once begun might be terminated sooner, if the participants improved their use of evidence. A knowledge of statistics is a tool for better understanding and mutual respect of each other's views. It is also a facilitator for changing one's own views as they become subject to the (scientific) tests of evidence.

While my own background is in economics, I have restricted the use of illustrations from my own field, as my invitation is not confined to economists; the examples are taken from all social science and arts subjects, including management and business studies. I have avoided footnotes and too many references and I hope that the reader, as consumer, approves. The bibliographical references are collected at the end of the book and give you an indication of the vast literature that is waiting for those who want to delve deeper than I intend to take them on this occasion. The invitation to write this *Invitation* was made to me by Michael Hay (of Martin Robertson) and he and his team were most helpful in their support during its gestation. I am also most grateful for the critical services of Professor Mark Casson, most of whose advice I took. Therefore the defects that remain are entirely the result of my refusal to accept even good advice when it conflicts with my vanity.

GK Edinburgh

1

The Great Debate

Before we can think about the purposes of statistics we must first think about thinking. Inevitably this requires a brief detour into a slightly more abstract, even philosophical, region. However, the detour is essential if we are to appreciate the essence of the statistical method.

A paradox

I begin with a statement from John Maynard Keynes: 'Part of our knowledge we obtain direct; and part by argument', he wrote. With this sentence, in his *Treatise on Probability* (1921), Keynes brought together two opposing views of philosophy. The debate between these views was all the more poignant because it was largely one that was confused by a fundamental flaw: neither school was right in their estimation of what was at stake! At the time the argument was, and to some extent still is, about which of the two sources of knowledge identified by Keynes is primary over the other. Do we get our ideas about the world exclusively through our experience, our senses, our observation and such like (that is through *induction*), or do we get them by conceiving in our minds rational and logical 'laws' of what must be happeneing *if* the world corresponds to our assumed interpretations of it (that is some form of *deduction*)?

This debate is of relevance to statistics not only because it represents a fundamental division of views about the source of knowledge, but also because it is against a largely philosophical background that statistical methods are judged as generators of knowledge. Also, an appreciation of the issues in the great debate between deductionists and inductionists has the useful benefit for students of statistics of identifying the modern role of statistics as a descriptive and inferential science.

Statistics is, above all, an empirical method. It collects, organizes, manipulates and 'interprets' the significance of numbers (known as *data*) from the real world. The rules under which it gathers and operates on this data cannot be isolated from the theories, and indeed the practice, of science.

Statistics, for example, does not deal with the truth or otherwise of thoughts and perceptions, except where they manifest themselves in the expressed opinions or actual behaviours of those who hold them. The statistician can count the numbers at prayer but he cannot judge the validity of the faith of those on their knees – though he might have a view about the connection between a specific act of prayer and the likely occurrence of the prayed-for event!

The statistician has to find a way of measuring relationships, sometimes from only a small sample of a larger number of actual cases, and sometimes between things that have no obvious quantifiable association. In addition, the statistician has to be satisfied that the results obtained will survive the scrutiny of others versed in statistical theory and techniques. To ignore this point exposes the statistician to the ridicule of contemporaries.

Though the techniques of statistics can be mathematically rigorous, the data only lends itself to inference rather than certainty. It is this parodox, of a methodology that is firmly rooted in certainties of mathematical theorems, and which is at the same time, of necessity (because of the nature of data), given only to uncertain inferences, that gives statistics both a popular (and unjust) aura as a discipline half-removed from deceit ('lies, damned lies and statistics', and so on) and gives its philosophical foundations the accolade of the nearest thing to empirical certainty that there is this side of omnipotence.

Practice and theory

To underline these points we begin with the early history of civilization. People could count long before they could theorize about counting, yet only through abstract thinking rather than practical application could civilization rise from merely arithmetical practice to mathematical theory.

The Egyptians and the Babylonians developed a sophisticated arithmetic and used it for several centuries in the practical daily life of their societies. An ability to measure space was a tremendous leap forward in the history of the human race but the Egyptians and Babylonians never developed a *theory* of spatial relationships. That was left to the Greeks.

The priests who marked out the squares and rectangles in the mud flats by the Nile and Euphrates rivers acquired a facility for accuracy by practice and experience; they learned by observation and by trial and error. Their achievements were magnificent, as is confirmed by every tourist who gapes (and all do) at the remnants of the once great pyramids at Gizi, but trial and error is a time consuming and limited method for discovery – a civilization could come and go before a necessary, because illusive, connection might be noted between one set of data and another. Independent civilizations might have to 're-invent the wheel', or its equivalent, in whatever passed for their scientific knowledge.

The Greeks were different. They started with the idea of space rather than the use of it. They were not so much interested in *the* triangle drawn across the field as they were in *all* triangles that exist, or could exist, in the universe. They were more interested in the universal triangle, square, sphere or solid than they were in any specific one that someone could draw *imperfectly* with crude instruments.

This was a different approach to that of the practical geometer – a mathematician was a thinker, not a doer. The priest, on the other hand, was interested in experimenting with his numbers only until he found a way of doing something – like measuring a field, or lining up an artifact with a heavenly body. Upon his ability to do this lay his earthly power and influence (and, indeed, his affluence).

The problem of induction

The experimenter's approach to the generation of knowledge is the basis of the philosophical method known as *induction*. A conclusion is induced from the evidence of repeated experiments. If a similar result is obtained each time it is tried, the priest concludes that the result will always be obtained.

Making assumptions from a limited experiment, however, is the weakness of the inductive method. For example, a child who touches a burning coal need not know anything about the theory of combustion to 'know' to avoid such contact in future. Any subsequent contact with burning coal reinforces the message in the child's brain to avoid anything that looks like burning coal. Consider now a child who runs across a road safely without looking carefully for traffic. This experience, if repeated, could be the basis of a highly risky life-style. It also exposes the limitations of induction as a method of thinking about the way the world operates.

Whereas we can be confident that it is most unlikely that contact with a burning coal would not cause a burning sensation to an inquisitive hand, it cannot similarly be asserted, with any confidence, that a child will not sooner or later come into (perhaps fatal) contact with a moving vehicle if the child continues to run across roads without looking carefully for traffic.

Inductive reasoning from experience to a universal truth cannot prove anything *for certain,* and this has always been of deep concern to those philosophers who sought certainty in their sources of the generation of knowledge. The Scottish philosopher, David Hume, expressed what has become known as the *problem of inductive reasoning*. Hume convincingly argued the proposition 'that the sun will not rise tomorrow is no less intelligible ... and implies no more contradiction than the affirmation that it will rise'. Why is this so? Because, no matter how many mornings that the sun has risen in the past, it cannot *logically* be established for certain that it will continue to do so in the future.

Moreover, this problem is compounded if there is, as yet, no experience of an event or state – you are, say, contemplating a change in the way the world is organized. In this case there can be no inductive reasoning that will have anything worthwhile to say about the proposed change because it has no experience to guide it. For example, inductive reasoning cannot establish the prudence, or imprudence, of establishing, or not establishing, a socialist or fascist Utopia. It is a matter of faith, but faith is no assurance of anything happening remotely like that which is anticipated by those who advocate or oppose the change.

This is the key to the enormous importance of the contribution of the Greeks, who turned (for whatever reason, and there is a debate about it) from induction to *deduction*.

Euclid's deductive geometry

Deduction, or abstract reasoning, is not based on evidence. It starts with an assumption, which if believed to be true (why is not relevant) it follows that the conclusions must be true as well, unless there is an *illogical* step in the reasoning.

One of the many intellectual achievements of the Greeks was to derive systems of logical argument known as *syllogisms* (for example 'All men are mortal, and Socrates is a man, therefore Socrates is mortal'). A system of logically deduced syllogisms is impervious to error, and, as a consequence, anybody trained in deductive reasoning will recognize arguments that are illogical in their construction and therefore in error (for example the syllogism: 'All Greeks know of Socrates, Caesar Augustus knows of Socrates, therefore Caesar Augustus is a Greek' is in error because the premiss 'all Greeks know' does not exlude the possibility of non-Greeks also knowing of Socrates and therefore the conclusion does not follow logically).

If a mode of reasoning could be established that was impervious to error, it followed that reason could establish truths impervious to contradiction by experience. You should note here that by truths in the context of syllogistic reasoning we mean *logical* truths, which are not necessarily the same as *factual* truths (for example the syllogism that 'All Europeans who explored America were atheists, Columbus explored America, therefore Columbus was an atheist' might be logically true but it is a factual truth that Columbus, although a European, was also a Christian).

It is a short step from the belief that logical truths could not be contradicted by experience, to the belief that a logical truth did not need the corroboration of experience, and what a short cut that represented in the methodology for discovering scientific truths! No more endless experimentation that resulted, at best, in uncertain truths – deductive reasoning could produce certain truths that were universally 'true'.

Thus, if the geometry of the triangle could be established logically and without error, then you had access to a universal triangle and need never rely on one crudely drawn on a slate or across some mud. Geometers did not even have to draw one at all to appreciate the beauty of its properties. The triangle (or any other geometric figure) created by pure reason was available for contemplation and manipulation without the risk that a 'maverick' triangle existed somewhere that would contradict the properties of one created by reason.

The geometer, Euclid, collected in his books the achievements of Greek mathematics that established the properties of triangles, and many other geometric shapes and solids, that manifestly were deductively true and therefore impervious to contradiction no matter in what circumstance they were contemplated. We now know, in Einstein's universe, that the certainties regarding the truths of Euclid's geometry are unfounded – triangles, lines, squares, solids, and space even, do not necessarily conform to the properties that were so rigorously deduced in Euclid's books. But at the time, and for many centuries later, Euclid reigned supreme.

To get a 'feel' for the deductive method in its purest form, *scan* (that is do not read) through a traditional geometry text. Euclid's book begins with 35 definitions (for example 'a point is that which has no parts, or which has no magnitude'), 3 postulates (for example 'that a straight line may be drawn from any one point to any other point') and 12 axioms (for example 'things which are equal to the same thing, are equal to each other'). Upon these foundations an entire geometry was developed.

The importance of axioms to the deductive method cannot be exaggerated, for without them the entire edifice crumbles. Axioms by definition are self-evident truths which are admitted without proof. Indeed, if an axiom has to be proved it is no longer an axiom. For example, there is no point in looking for actual things that are equal to each other in order to establish an axiom of equality, for the axiom must be true in all and every case irrespective of experience: axioms are immune to experience, they are true by assertion. From axioms the theorems are deduced. They admit no room for error, and many hours were once spent by school children learning Euclid's propositions until they were word-perfect.

Newton is alleged to have remarked that he saw little point in Euclid working out all his demonstrations and proofs once he stated the axioms, because, Newton argued, the proofs were implicit in the axioms and therefore were inevitable once the axioms were stated. How many generations of school children wished privately that Euclid had come to the same conclusion before compiling his books?

On the basis of deductive reasoning, Greek science and philosophy flourished and it discovered much more about the world than any previous civilization. These stupendous achievements are reflected in the fact that much of Greek mathematics and philosophy has survived into our own civilization, as has much of its literature, poetry and theatre.

If deduction was worshiped by the Greeks, empirical methods were disdained, as was practical work of any kind, especially commerce. The Greeks turned up their classical noses at even the idea that experience could be a guide to anything worthwhile, and this was the seed of their undoing. Having mastered the art of being deductively brilliant they then fell victim to the ravages of a practical force which was utterly unsentimental and devoid of any intellectual understanding of the abstract. The military force that overthrew them was totally imbued with the successful disposition of practical power: the age of the Romans arrived and with it came destruction.

Nothing could more economically sum up the difference between the two great Mediterranean cultures than the murder by a Roman soldier of Archimedes who was studying a geometric diagram during the final stand of the defenders of Syracuse: for the next thousand years 'no Roman ever lost his life absorbed in the contemplation of an abstract diagram' wrote the mathematical philosopher, A. N. Whitehead.

The Roman age eventually passed, but this did not immediately renew the debate between deduction or induction as methods of scientific enquiry. There were several more centuries to go before the debate got underway again, and the change when it came is known as the Renaissance. Intellects gradually rediscovered the scientific method and the old authorities gave way, slowly, to new ideas about how knowledge progresses. The Great Debate recommenced.

Our concern, however, is not to follow the birth-pangs of the new scientific order. We are interested in how the great debate between deduction and induction recommenced and what form it took in the centuries to our own.

Bacon's new induction

Francis Bacon (1561–1626) was what today we would call a cross between a politician and a civil servant. His family was well connected at court, and Bacon rapidly advanced once he found favour with the Scottish King James VI (after a disappointing start with the English Queen Elizabeth). He rose to become Attorney General and finally Lord Chancellor, though he fell out of favour (under a cloud of bribery charges) and died with his political reputation in shreds.

Bacon challenged the accepted orthodoxy of Aristotle's great work, *Organum,* in a book he entitled *Novuum Organum* (1620). It brought into the open the contest between experience and *a priori* or deductive reasoning. Bacon chastized Aristotle for 'dragging experience along as a captive constrained to accommodate herself to his decisions, so that he is even more to be blamed than his modern followers, the scholastics, who have deserted her altogether'.

Bacon was as critical of experiments without thought as he was of thought without experiments. He noted that those 'axioms now in use' were not subject to the test of experience and also that when a contrary fact appeared 'the axiom was rescued and preserved by some frivolous distinction', whereas the truer course would be 'to correct the axiom itself'.

You might care to note that this tendency to *ad hoc* adjustment of theories when faced with damaging counter-evidence survives today – people can become emotionally attached to their theories, having invested perhaps years of work on them, and they are not inclined to abandon them 'just' because evidence begins to cumulate that there is something wrong with them. It is far easier to 'adjust' the theory to accommodate the evidence, where it is not possible to discredit it – or, better still, the person producing it!

There are, according to Bacon, only two ways of establishing the truth: there is deduction based upon general axioms which are supposed to be of indisputable truth, like the axioms of Euclidean geometry, and there is his as yet 'untried' method which 'constructs its generalisations from the particulars of sense', that is experience. This is an inductive method, as can be seen in the claim that it leads 'men to the particulars themselves, and their series and order' and which requires men to lay their preconceived 'notions by' and 'familiarise themselves with facts'.

Bacon recognized that facts by themselves were not enough – he was not an ultra-empiricist. 'Experience,' he wrote, 'when it wanders in its own track is mere groping in the dark, and confounds men rather than instructs them'.

Bacon recommended something bearing a remarkable resemblance to the hypothetico–deductive method of the late nineteenth century. He was aware that a single contrary instance exposed an hypothesis 'to peril', and he sought to use the power of a negative instance as a means of refining a theory until it could stand the test of as many instances as were practicable, that is until affirmative instances, rather than an axiom, suggested the truth of the hypothesis. Thus, he sought to deduce 'causes and axioms from effects and experiments; and new effects and experiments from those causes and axioms', and in so doing, to uncover 'nature's forms' (that is causes).

Bacon recognized that induction would require greater labour 'than has hitherto been spent on the syllogism'. Bacon's inductive method, whatever its ultimate fate in the history of science, did establish, at least in the written record if not in the minds of scientists, that the weakness of induction – a single contrary instance discredits an assertion – could be turned on its head – the absence of contrary instances while not proving a theory true would at least align it closer with the laws of nature, and in doing so could expose nature's laws to human contemplation.

Descartes and the resurrection of deduction

René Descartes (1596–1650) had an altogether less orthodox career than Bacon. He was a catholic, yet as a mercenary sold his military services to the protestant cause in the Thirty Years War. He became a philosopher and mathematician as a result of 'visions' he had while on campaigns, and he settled in the Netherlands to write treatises and, hopefully, to avoid controversy. His last service was to the Queen of Sweden who appointed him because of his reputation in Europe. However, she did not keep her palace warm enough for Descartes and he died of pneumonia in 1650.

Rene Descartes took a different road from Bacon to the truth. He resurrected the deductive method. Descartes, literally, had a dream (or two dreams according to some accounts). It may have been the after-effects of adrenalin depletion while soldiering, it may have been a too full or a too empty stomach, or even a touch of *vino veritas,* but whatever it was, during the night of 10 November 1619, Descartes revealed to himself, in a dream, that the universe had a mathematical structure.

The impact of this revelation was to shape his life and modern mathematics. If the universe had a mathematical structure it followed that the structure could be understood by the study of mathematics. The geometers had shown the mathematical way to discover truths and the same methods could produce a 'universal mathematics' applicable to all knowledge: hence, purify human reason with universal axioms and deduce truthful theorems from the axioms.

Descartes decided that if he could eliminate doubt from any proposition he asserted in his mind then that proposition must be true. His famous assertion *cogito ergo sum* (I think therefore I am) was an example of a proposition from a doubt-free mind. He felt that 'all the objects that had ever entered into my mind when awake, had in them no more truth than the illusions of my dreams' and excluded them from his catalogue of truths. With this draconian step he eliminated sensory experience from his method and took off in an entirely different direction to that of Bacon (though elsewhere in the *Discourse* he concedes that some experimentation may be necessary due to the imperfectability of man's powers of reason, but then only to show, and not to be, the way).

He also prescribed that an exhaustive working and reworking of the steps in a deductive reasoning was necessary to purge it of error, much in the manner of establishing a Euclidean proposition. In fact, the entire approach bears comparison with the Euclidean method of establishing a geometrical proposition.

As for his philosophy, it produced a deductive 'proof' of the existence of Descartes himself *(cogito ergo sum)* and also of the existence of God (which 'is at least as certain . . . as any demonstration of geometry can be'). But important as Descartes was to the renewal of the great debate in philosophy (and to mathematics, for which he created the algebra of geometry) essentially his was a dead-end for philosophy. Reasoning, rigorous and productive as it could be, could not ignore the requirement that its conclusions, if they were to have any practical value, ultimately had to conform to experience.

Mill's inductive logic

Returning to the problem of induction – that no experience can eliminate the possibility that a similar experience will or will not occur in future – we can note that this remained an irritating flaw in the increasingly more sophisticated philosophical discourses of the nineteeth century. It was not to be resolved satisfactorily, yet many brilliant attempts were made to do so.

Among these was John Stuart Mill in his *System of Logic* (1843). Mill was the product of a terrifying education programme imposed on him as a young child by his father, which today would excite the intervention of the authorities. At three years old he was made to learn Greek (classical not tourist). Before he was twelve he had also learned Latin and read the Classics from Plato to Aristophanes by way of Horace, Ovid, Virgil and many others. He also triumphed in geometry, calculus, logic and political economy, and all this before his contemporaries went up to university.

Fortunately, his was a prodigious intellect that rested on an appropriate if strange personality – the kind of affectionless and dedicated education that was imposed upon him by a determined father would have collapsed most other people into insanity or suicide, or both. Even for Mill it was a close-run thing. At age 26 'it occured to me', he reveals in his *Autobiography,*

> to put the question directly to myself: 'Suppose that all your objects in life were realised; that all the changes in institutions and opinions which you are looking forward to, could be completely effected at this very instant: would this be a great joy and happiness to you?' And an irrepressible self-consciousness directly answered, 'No!' At this my heart sank within me: the whole foundation on which my life was constructed fell down.

Mill eventually recovered from his acute depression and went on to write long, detailed and authoritative works, published in many editions, ensuring his domination of the liberal reforming strand of Queen Victoria's Britain. The social mores of the age suited him well and, like a character in a Victorian novel, he met, fell in love with, cherished and courted a married woman, Harriet Taylor, apparently waiting for twenty years, until her husband obligingly passed away, to consumate virtuously their love in wedlock. From Harriet he learned a lot, not least about what today we know as feminism, and when she died he turned his innocent affections (bordering on adulation) towards her daughter.

Mill in his *System of Logic* (1843) built on Bacon's work. He rejected mere enumeration as a method of induction, exposing its fallacy in the supposed belief of people in Central Africa that the entire human race was black, or in the supposed inductive reasoning of Europeans that all swans were white. In both cases the false conclusion followed from the lack of evidence to the contrary, until Europeans visited Africa and Australia.

The difference between a simple enumerator and the 'superior mind' is that the former is merely a passive observer who 'accepts the facts that present themselves, without taking the trouble to search for more' while the latter 'asks itself what facts are needed to enable it to come to a safe conclusion, and then looks out for these'. Modern induction, according to Mill, required a search for what David Hume called antecedent causes; those that are 'discoverable not by reason but by experience', and he set out five canons of induction which would achieve this object. He distinguished between observation and experiment by noting that observation can identify 'sequences and co-existences, but cannot prove causation' while experiment can do all that and more; it can experiment on either the causes or the consequences of a supposed relationship by devising artifical circumstances in which tests are made of both.

Having established rules for experimental science, Mill had to face the problem of applicability of such methods to those sciences that did not provide a facility for experiment, such as astronomy, politics, political economy and history. Society, Mill argues, is much more complex than simple physical phenomena, like gravity and light, and therefore to attempt to use isolated examples as evidence for or against a particular combination of events is patently ludicrous because 'it is always certain that before the effect of the new cause becomes conspicuous enough to be subject to induction, so many of the other influencing circumstances will have changed as to vitiate the experiment'. This leaves a wide area of human experience (especially in the social sciences) beyond reach of the inductive method. Mill suggested that another method 'which considers the causes separately, and infers the effect from the balance and the different tendencies which produce it; in short, the deductive or *a priori* method' is appropriate in these circumstances.

Having conceded that induction is inappropriate for making statements about the social sciences (and astronomy), and that deduction is all that is available, Mill does not entirely abandon induction altogether. He insists that deductive reasoning must begin with an inductive premiss, that is some idea of the supposed relationship between cause and event that is founded in inductive evidence, which '(pre)supposes a previous process of observation and experiment'.

In this, Mill is driven to incorporate deductive reasoning into his method by the patent inapplicability in the social sciences of experiments and the impossibility of separating the influence of multitudinous causes from each other on effects. He asks in respect of the events which are the concern of social science:

> When in every single instance a multitude, often an unknown multitude, of agencies, are clashing and combining, what security have we that in our computation *a priori* we have taken all these into our reckoning? How many must we not generally be ignorant of? Among those that we know, how probable that some have been overlooked; and even were all included, how vain the pretence of summing up the effects of many causes, unless we know accurately the numerical law of each – a condition in most cases not to be fulfilled; and even when it is fulfilled, to make the calculation transcend, in any but very simple cases, the utmost power of mathematical science with all its modern improvements.

With this statement, Mill almost knocks at the door of statistical inference. He is concerned with the impossibility of certainty, both in the identification of all causes at work in a particular instance, and in the relative importance of these causes as determinants of the outcome.

Instead of going on to make statistical inference a guide for our judgements, Mill turns to what is now known as the hypotheto-deductive method. In this approach, the scientist states a hypothesis about a relationship, and subjects the conclusions that are deduced from that hypothesis to the test of experience: 'these conclusions must be found, on careful comparison, to accord with the results of direct observation wherever it can be had'.

In this view, if the conclusions are found to accord with experience, the hypothesis is *verified*. The weakness of verifing conclusions on the basis of careful comparison with direct observation is similar to the problem of induction generally – how do we know that a successful comparison will not be followed by an unsuccessful one? In other words, can we be sure that we will not find a black swan somewhere someday?

Modern views have inverted the verificationist approach: if

the hypothesis is not contradicted it remains a conjectured explanation of the alleged relationship; if the hypothesis is contradicted, in even, a single instance it is rejected. Thus, no amount of experience can confirm an hypothesis; it can never be *verified* or proven to be true by experience. The evidence can, however, reject it decisively.

Mill states the verificationist method most clearly:

> An hypothesis is any supposition which we make (either without actual evidence, or on evidence avowedly insufficient) in order to endeavour to deduce from it conclusions in accordance with facts that are known to be real; under the idea that if the conclusions to which the hypothesis leads are known truths, the hypothesis itself either must be, or at least is likely to be, true.

Note the use here of tentative words like 'probably', 'likely', 'believed to be' and such like. This is most important, for it suggests that the deductive–inductive debate was shifting from a search for a system that would produce certain knowledge to one that could, in principle, provide acceptable levels of confidence short of certainty.

In this context it is highly significant that chapters 17 and 18 of Mills *System of Logic* (8th edition, 1925) are entitled: 'Chance and its elimination' and 'Of the calculation of chances'.

Uncertainty and social science

The hypothetico-deductive method with its verificationist overtones dominated scientific methodology by the end of the nineteenth century. Induction had to give way to deduction in the form of stating some idea or hypothesis of the likely relevance of particular sets of empirical data. Scientists do not trawl randomly through data in the hope that something will turn up; they have to start with some notion, however vague, of where to look and what to look for. This vague notion was called an hypothesis which did not originate from evidence – it was a deduction.

The situation with deduction was no less compromising.

Deduction from axioms isolated from experience – indeed, in some cases directly contrary to facts – could produce impressive conclusions, some of them relevant (for example, in mathematics and pure physics), but at some point the evidence of the real world is decisive for the deduced conclusions. A science totally isolated from experience conceivably has a role to play, but mainly as a signpost to empirical study. Deduction tells observers and experimenters what to look for and where to look for it, and somebody sooner or later looks and reports on what is found.

The growth of social science – mainly, but not exclusively, in the field of political economy – at the end of the nineteenth century was itself a contributory factor to the decline of inductive reasoning. Deductive reasoning from axioms about human behaviour – economic, political, sociological, psychological – were commonplace. *Homo Economicus,* sometimes in the person of Robinson Crusoe, strode across the pages of books on political economy, and considerable progress was made in the axiomatic analysis of economic behaviour as a result.

For social scientists, the lack of certainty, the extremely complex nature of the phenomena that is studied, the difficulty of making meaningful experiments and the relatively non-quantifiable nature of much of the interactions that are considered to be relevant, is no longer a criterion for demarcating between natural sciences (the so-called 'proper' sciences) and social sciences (so-called 'pseudo' sciences).

If uncertainty is no longer a disqualification for making meaningful statements about the world, we are at once liberated from being stymied by imperfections in our data, assuming we can find rigorous ways of collecting, presenting and analysing the evidence that is available in order to make less than certain, but nevertheless probable, inferences from it.

And what of the role of evidence in our scientific work? This raises a number of questions which the rest of this book addresses, such as:

What evidence is relevant in deciding on the relative reliability of an hypothesis?

What evidence can be considered to be decisive in rejecting a given hypothesis?

What happens when the evidence is mixed, some for and some against the hypothesis?

How can we protect our scientific integrity and only work with 'proper' experimental methods that genuinely amount to a sufficient test of our hypotheses?

The search for answers to these types of questions boils down to arriving at agreement on a 'proper' methodology for all social scientists to work with, which will provide a means for assessing the evidence for and against the alternative and contrary hypotheses that we are studying.

If we can agree on a code of scientific conduct for making judgements about hypothetical statements we will have progressed a long way towards establishing the nearest we might get to 'objective' criteria for making judgements in social science. This book is about such a common scientific methodology, namely that of of statistics. This is not to say that the statistical techniques discussed in the pages that follow, and their use, are immune to criticism or controversy. Far from it; any technique is open to challenge for being inappropriate to the problem in hand.

An ability to tamper with evidence is not confined to statistics or statisticians (though judging from 'popular' witticisms about statistics there is a general prejudice against it: 'figures can lie, and liars figure' for example). But verbal reasoning too can be full of cheating tricks. Ironically, a major difference between statistics and verbal reasoning is that with statistics the 'rules' of the method are widely known among scientists and it is very difficult to 'cheat' on the evidence without being 'found out' by colleagues and critics. The same cannot be said for the gullible public, or for that matter equally gullible social scientists, who have avoided learning anything about statistical techniques and rely entirely on verbal reasoning.

Statistical inference as a technique was not an overnight revelation that occurred in step with the philosophical debates discussed in this chapter. For a long time statistical inference developed its techniques and refined its methods somewhat isolated from philosophical debates on the sources of knowledge and the methods of acquiring it.

Much that is important in statistical inference was not developed until the twentieth century. But when it was needed a reasonably well worked out, if as yet underdeveloped, methodology was available. As it spread into the social sciences it took on a new lease of life and gained a rapid momentum. The age of a statistical social science arrived.

2

From Counting to Statistics

The recognition that measurement was a legitimate and necessary activity in the study of society did not gain wide acceptance overnight. Historically it took a long time from the emergence of the practice of collecting numerical data to the first attempts at making inferences from it; statistics is a comparatively recent discipline.

History, unfortunately for those who prefer their lives to be tidy, is a very untidy process, and there are the usual academic disputes about the date of birth of statistics.

Early traces of the use of numbers

The fourth book of Moses, appropriately, is entitled 'Numbers'. It describes the taking of a census of the people of Israel, or rather, those of them that were thought worth counting, namely, all males 'from twenty years old and upward'. Women and children were not to be counted (and therefore were not – Moses being careful never to offend his God):

> 'As the Lord commanded Moses, so he numbered them in the wilderness of Sinai' and he found in the twelve tribes a total of 603,550 males, aged 20 and upwards, who were 'able to go forth to war in Israel' (Numbers, 1, v. 19, 46).

Counting people was known in history even before the time of Moses and it has been practised from time to time over the centuries since by Kings and governments, usually to assess the numbers available for military duty or taxation.

For example, we have reference in the New Testament to a famous Roman census:

> Now it came to pass in those days, there went out a decree from Caesar Augustus that all the world should be enrolled. This was the first enrollment made when Quirinius was governor of Syria. And all went to enrol themselves, every one to his own city (Luke, 2, v.1 – 3).

Like the numbering undertaken by Moses, this cannot be thought of as an early example of a *statistical* approach. Something more is needed beside the passive collection of numbers, for numbers by themselves are like sleeping giants whose inherent power is untapped. This applies to the famous, and fascinating, *Domesday* Book of William the Conquerer, and to Charlemagne, who had his possessions counted, a couple of centuries before the *Domesday* Book.

There is abundant evidence of counting, and accounting, in the non-statistical sense throughout history. For example, I have a copy of *Papers Relating to the Army of the Solemn League and Covenant 1643*–47 (2 vols., Terry, 1917)which details the disbursements to the army's officers and men, and even to their widows and children too, as well as the price and quantity paid for every single item received to conduct its wars. This material is fascinating to those interested in the religious (bad) temper of seventeenth-century Scotland, or the Cromwellian wars in England, but records, even those that are abundant in detail, do not constitute evidence of statistical thinking.

It is a common modern practice to loosely describe data, almost any data, as statistics. We refer to a country's 'vital statistics' which describe the condition of a country at a specific time by detailing an inventory of its incomes, possessions, roads, rail, and property, and so on. Strictly, such data only imply the presence of a statistical approach, they do not confirm it. For everyday purposes we can safely think of our modern collections of data as statistics, indeed that is how they are described in their titles and by those who quote them, only because we know that there is a statistical purpose behind their collection.

Now that we live in the age of statistics we must be careful we do not read too much into the historical evidence when we are searching for the origins of statistics as a method. When Robinson Crusoe first saw footprints in the sand he assumed that his island was inhabited by others, because living people leave footprints, only to find, to his chagrin, that they were his own footsteps!

It is essential, therefore, to distinguish between counting and statistics. Mere enumeration of things or people is *not* yet statistics though an ability to do so is a necessary preliminary step if a science of statistics is to become possible. Something more is needed besides the summation of numbers. There has to be an element of *interpretation* of the data before we can say that a statistical operation has occurred, and if we are to identify correctly when statistics, as opposed to numbering, was discovered we must search for the first appearance of an attempt at intepretation of what the numbers suggested (whether it succeeded is of no matter).

On this basis, we can date the origins of the statistical approach as the middle of the seventeenth century, roughly with the publication of a book written by a London draper, *Natural and Political Observations Upon the Bills of Mortality,* in 1662.

The two aspects of statistics

To some extent, the problem of the origins of statistics is complicated by the fact that we talk about statistics in two fairly distinct ways, though together they constitute its subject matter. The two ways are distinguished from each other both in purpose and methodology and each can be traced back into ancient history.

Roughly, statistics can be either *descriptive* or *inferential.* Descriptive statistics are what the name implies – they are used to describe something, but in a special and organized way:

> Descriptive statistics are about presenting in some convenient way, either by number or graph, some noteworthy features of available data.

There is an inferential element even in the purely descriptive approach – the data may lend itself to an 'obvious' interpretation, such as whether the graph line is rising or falling, but the inferential element is normally extremely limited, being confined to what is easily discernible.

The statistical techniques that present whatever is found to be worthy of note, are fairly well established and normally constitute the first part of a basic statistics course. It is also the case that most popular beliefs (and specialist's complaints) about the abuse of statistics originate in the mis- or non-use of these fairly elementary techniques. This is one reason why students are expected to become familiar with elementary descriptive techniques and also with the more obvious of the common frauds perpetrated on the innocent.

Inferential statistics is about making inferences, or forming relatively sound opinions, about something in a systematic and rigorous way:

> Inferential statistics are used to make meaningful statements about noteworthy features of events that have not been directly observed or have only been partially observed.

Generally, inferential statistics use more sophisticated techniques than the purely descriptive ones, which, for the interested specialist is a (harmless) delight in itself. However, the delight, or rather its consequences, ceases to be harmless if it leads the specialist to assume that the sophistication of a statistical technique ensures the reliability of the inferences it provides. This is similar to the unsafe assumption of consumers who judge the quality of a good by its price.

In addition, the increased sophistication of the inferential statistic raises the level of difficulty involved in understanding their operation and meaning, which can turn off the marginally interested student.

This chapter is largely about the development of descriptive statistics in sociology and economics but of necessity it strays into the often parallel development of inferential statistics, which is taken up in greater depth later.

The world's first statistician

John Graunt (1620–1674) was a tradesman draper. He walked in the shadows of what passed for the great and good of Restoration England and his very ordinary background and occupation has tempted some authors to question his sole claim to fame (but alas not to fortune, for he died a bankrupt) as the author of the first ever book on statistics, and transfer it to his friend, Sir William Petty (1623–1687). It is bad enough for a scholar to have his reputation questioned while alive, but to have it questioned after his death is doubly onerous as he has no right of reply.

Graunt lost his property and source of income in the Great Fire of London and, more tragically, given that one can recover from material misfortune, he made the tactless 'mistake' of converting to Catholicism. When he most needed the earthly help of those able to provide it, his spiritual beliefs gave them an excuse not to do so. However, the precise authorship of the *Observations* is probably of less importance to statistics than the fact that somebody collected the data and attempted a statistical interpretation.

The material for Graunt's observations had first to be collected by someone, and the motivation for making such a collection was the simple one of the preservation of the King's health. London had been swept by a particularly nasty bout of plague in 1603 and following this it was decided to keep a weekly watch for a rise in plague deaths in the city as an early warning for the King to decamp his capital and seek the relative safety of his countryside.

Graunt states in his preface that the collectors of the mortality figures 'made little use of them, than to look at the foot, how the burials increased, or decreased; and among the casualties, what had happened rare, and extraordinary in the week'. Making 'little use' of numbers is a sign of a pre-statistical arithmetic. Until use is made of the numbers they remain dormant.

The compilers of the Bills of Mortality were a mixed bunch of untrained 'searchers', and little reliance could be placed on the accuracy of their classifications, particularly where they were left to judge the cause of a person's death. There were other weaknesses in the data. For instance, though they recorded the christenings in their parish, this obviously was not the total of all births because not all citizens were members of the established church nor were all children christened in it.

On the basis of the available data, Graunt made an estimate of the population of London. This is our evidence of the first interpretation of passive data and marks the beginnings of what we now call statistics.

That Graunt made reasoned estimates on the basis of a specified method separates him from all the enumerators of the past centuries. By making estimates I do not mean he made arbitrary guesses about the basic data – such guesses have always been (and still are) practised among those who know little of the importance of their work. Graunt's methods of estimation were reasonable inferences from the evidence, and being specified they were subject to amendment by improved inferences – 'good reasons must, of force, give place to better'!

Graunt knew the number of births, or rather an approximate figure for them, allowing for the minority of catholics and non-conformists in the population. He estimated the number of women of child-bearing age by suggesting that this should be about double the number of births 'forasamuch as such women, one with another, have scarce more than one child in two years'. Modern women would no doubt be horrified at a pregnancy every two years for 24 years but Graunt lived in seventeenth-century England. From this he supposed that the number of families would be about twice the number of women of child-bearing age on the grounds that the women who were married (aged between 16 and 76) would total twice the number of those between the child-bearing ages of 16 and 40.

Graunt reflects the social mores of his day – and living close to the people as a trader he is likely to have had accurate notions about what constituted a representative family. He estimated that each family would consist of eight persons on average – the husband and wife, three children, and three servants or lodgers. It was then a simple matter of multiplying up from the number of 12,000 recorded christenings:

12,000 x 2 = number of fertile women = 24,000

24,000 x 2 = number of families = 48,000

48,000 x 8 = total population London =384,000

There are undoubted errors in his method but again this is not important. It is the discovery or demonstration of a statistical approach that is the crucial event from the historical point of view. In an open society such as Restoration England, once the statistical method was made public it inspired others to imitate and to improve on this first crude attempt.

If you felt uneasy at all at the assumptions in Graunt's calculation of the population of London, then you are already thinking in a statistical way, though you may not yet appreciate the significance of doing so.

Science is about asking questions – and while much of scientific progress is about asking the *right* questions, the habit of asking any questions at all is an important characteristic of its method. But science is about more than just asking questions, *it is also about questioning the answers.*

The most important visible characteristic of a statistical approach to quantitative data is the way in which you are led to question the answers, and, behind the answers, the methods of deriving them:

Is that a reasonable assumption to make?

Why does she assume this is a representative sample?

How does he know his data is reliable?

What checks are there on the enumeration?

Is her arithmetic right?

Does that answer sound about right?

Is my surprise at the answer anything to do with my attitude to its implications?

Can that experiment be replicated to check on his conclusions?

Can that method be used in these circumstances?

and so on.

The first sample

Graunt's genius opened up the power latent in numbers for human consumption, leading eventually to today's discipline of statistics, with its established methods for attempting to answer the above types of questions and myriads more. We can see a little of how far he got by looking at some of the other calculations in his book.

There is a lot in Graunt's little classic. From past data in the Bills of Mortality he was able to estimate (using what today we would call a *time-series)* the growth of the city's population and, using this information, show the remarkable recovery within the space of two years in the city's total population after the worst ravages of the periodic plagues (he deduces this from the return to the average number of christenings of the pre-plague years).

This is another example of his drawing a significant conclusion from the raw data, as was his conclusion that the upsurge in the city's population could not have been caused by the natural birth rate of the inhabitants (as was popularly believed) but must have been caused by immigration because even on the basis of his overestimate of female fertility (the actual fertiliy rate was about half of his estimates) it was not possible for the luckless child-bearing women to have produced enough children to reach the population levels that he estimated existed.

Most significantly, to check on his calculations, Graunt went to the evidence. He took three parishes as *representative samples* – perhaps his most important innovation – and counted the number of families in them and the number of burials per family (three deaths per eleven families). With this he mutiplied the number of deaths in the Bills of Mortality by 11/3, giving him 47,666 or pretty close to his estimate of 48,000 families in the city (if you have a calculator why not check his arithmetic?

He supported this sample method with another. He used a map of London to estimate the density of families within a hundred yards square (he worked it out at 54 families), and as the city within the walls had 220 squares (he got that from a map), he calculated that this gave 11,880 families in all. Next, using the deaths within the walls (3,200) from the Bills of

Mortality, he was able to say that the familes within the walls were one quarter of the families in London, because 3,200 was a quarter of the 13,000 deaths per year. If his estimate of 11,880 families within the walls of the city is multiplied by four we get 47,520 families for the whole of London. Again this corresponds closely to his deductive estimate of 48,000 families on the basis of multiplying up from the number of fertile women and the birth rate. Thus Graunt felt confident of his figures.

Graunt also noted a slight majority in favour of boys over the number of girls christened over the years. His reaction to this illustrates the statistical approach once again, though, unhappily on this occasion, negatively. His conclusions about the relatively stable majority of boys born exhibited, not for the last time, that statisticians can come to wrong conclusions on the basis of their inferences from data: no statistician is ever immune to fallacious conclusions on the basis of unwarranted assumptions. Graunt writes:

> We have hitherto said, there are more males than females; we say next that the one exceed the other by about a thirteenth part. So that although more men die violent deaths than women, that is, more are slain in wars, killed by mischance, drowned at sea and die by the hand of justice; moreover more men go to the colonies and travel in foreign parts than women; and lastly, more remain unmarried than of women as fellows of colleges, and apprentices above eighteen, etc. yet the said thirteenth part difference bringeth the business but to such a pass, that every women may have an husband, without the allowance of polygamy.

Graunt's conclusions were forced from the data. He makes the mistake of searching for a 'reasonable explanation' of the facts on the basis of what sounds plausible, and while this is a useful way to generate hypotheses we must remember that our explanations (especially those that suit our prejudices) ought also to conform to the evidence.

It is also worth noting here that people often reverse the process and seek to 'prove' their prejudices on the basis of interpreting their facts. John Arbuthnott, using similar data, but for a much longer period (1629 – 1710), came to the conclusion that:

> Polygamy is contrary to the Law of Nature and Justice, and to the propagation of the Human Race; for where Males and females are in equal number, if one Man takes Twenty Wives, Nineteen Men must live in Celibacy, which is repugnant to the Design of Nature; nor is it probable that Twenty Women will be so well impregnated by one Man as by Twenty.

The hostility of authorities to polygamy and their favouring monogamy was, of course, partly a religious prejudice against Islam. In both Graunt's and Arbuthnott's cases, however, we have an example of the noting of a regularity in the data and it is this aspect which is important and not the singular and curious conclusions that they might have drawn. Noting regularities, and searching for them, is a characteristic of the statistical method and much of descriptive statistics concerns techniques for discovering if the raw data has any regularities worth noting.

But Graunt made another contribution to statistical methodology that towers, in some respects, even above these enormous strides forward: he made an estimate of the life table of a population of 100 people, that is, the proportion of the 100 people who would die each year from birth to aged 70 plus.

The Bills of Mortality did not provide information on the ages of the persons who died in a week and without this information it is not obvious how a life table could be drawn up, even if one was conceived. This type of problem is a test of the statistical method: can we find a way of inferring noteworthy conclusions from apparently unhelpful data? Graunt's genius was shown in his attempt to do so.

To make progress he realized that he had to approach the problem indirectly if he was to infer something that was not explicitly stated in the figures. We can see his reasoning from what he did and it is ingenious.

First, he counted how many causes of death, as listed in the Bills of Mortality by the searchers, applied exclusively to children and not to adults. This gave him the number of children that had died that year of children's diseases and disabilities, but it could not be the complete figure of deaths among children because some children would also die of diseases or disabilities that would also kill adults. Therefore, he had to estimate the proportion of children that had died of 'adult' causes of death and add these to his first figure to get the likely total of all children who had died each week.

Graunt separated out those causes of death which would apply to children of up to five years only ('Thrush, Convulsions, Rickets, Teeth and Worms, Abortives, Chrysomes, Infants Livergrown and Overlaid') and added to this an estimate of the proportion of death by diseases common in adults (in this case he chose to assume that about half of all deaths due to the collectors of the mortality data ascribed 'age' to 7 per cent of the weekly death toll and he assumed (tempting providence, so to speak) that the rate of death was relatively uniform throughout the years between 6 and 76. This is a clear error, because the same proportion of people in each age group do not die each year, but the error was not one that was obvious to Graunt (or anybody else) given the paucity of data that was available. But again, being a newcomer to statistics, you should note that having made public his assumption, others could improve upon it (and, in fact they did, as we shall note later).

How Graunt took the next step and calculated his famous Life Table is not known because he did not say, and typically it is the source of some controversy in the literature. However, his table, the first ever life table published, is given below.

Age	*Survivors*
0	100
6	64
16	40
26	25
36	16
46	10
56	6
66	3
76	1

The fact that it is quite wrong is neither here nor there; that it was produced by reasoning from raw data gives it its statistical importance. Graunt's attempt at a life table began a process which eventually led to the profitable foundation of a life insurance industry. This required the accumulation of much more data and an improvement in the method of deducing more accurate survival rates from the data, but before this could commence the path had to be beaten by someone.

The picture then is fairly clear: Graunt's little book started a small industry that gradually gathered momentum and eventually became commonplace all over Europe and North America. By the early nineteenth century, statistical records were primitive. But that is not the point. Graunt had created the *possibility* for statistical inferences to be made from a data base, and for the methods of doing so to become a common currency of comment and debate.

True, it was not yet clear what the full extent of the latent power of numbers was, nor were the techniques of tapping their power all that sophisticated, nor were the explanations for the numbers that emerged all that scientific (in some cases they were anti-scientific). Statistics had a long way to go, but with Graunt's first step, the long journey at last could begin.

In summary, because of Graunt's work, 'mere' numbers would never be 'mere' again.

3

The Triumph of the Numbers

After Graunt, we can fairly say, came the trickle and then the flood. Throughout the next two centuries, the enumeration of vital statistics, and their use in statistical operations, gradually increased to a torrent, such that by the end of the nineteenth century there was a considerable amount of data available and more was steadily being collected all over the world.

It is essential, as we have noted several times, to have numbers before we can make statistical inferences. Graunt's genius was in showing what could be done with data, even data collected in very trying circumstances and with very inadequately trained labour. It was inevitable, therefore, that the new discipline of statistics should go through a period of preparation during which the almost total lack of data would gradually be transformed into one where the collection and storage of data was a commonplace activity of the state. As with many other things, the process by which we achieve what is obviously in the best interests of those concerned is seldom a smooth one, and it has taken several centuries to establish data bases in which we can have much confidence. In this chapter we look at some of the pioneers and the significance of their work.

Petty's proposals

Sir William Petty, in my view, is a trifle enigmatic. His reputation has survived but one gets a certain feeling about the man that leaves a doubt or two about his 'soundness' – his speculations often rashly ran way ahead of the ability of his empirical data to support them. This is not to say that the man did not deserve honour; he is credited by some authorities with founding political economy and also with contributing enormously to the development of statistics.

Petty was a friend and confidant of Graunt. He also knew everybody who was important, and rose from cabin boy to knight. To sum him up in a sentence (grossly unfair of course): he was bright, enthusiastic, charming and, above all else, he was a talented opportunist, especially where it came to making money out of public office. At age 20 his worldly wealth amounted to £70, yet he died worth (on his own account) about £15,000 a year, which in those days made him extremely, nay, grossly, well off.

He made much of his wealth in the misery of the carve-up of Ireland following Cromwell's invasion and the English occupation. The restored Charles II knighted him, for what is not clear, though Greenwood (1941) suggests it was because a sovereign, on this occasion, *was* amused.

Petty was a proposer of vast schemes, one of which attracted the interest, if not the patronage, of a king whose expenditures always threatened to outrun his income: Petty suggested that it was possible to raise the taxation of the country without endangering its prosperity–music, no doubt, to a king's ears.

Petty proposed the establishment of a central statistical office that would collect details of births, marriages, deaths, ages and gender of the people, plus details of their houses, servants, number of hearths (taxable in his day), their land holdings, revenues, education and occupations. In addition, the figures were to be sorted into useful groupings by age, status and gender.

The evident benefits of such a register for a state were less obvious in the seventeenth century than they were to become by the nineteenth century and the proposal was not implemented for 150 years.

Petty grasped the significance of the life table first suggested by his friend, Graunt, and he clearly intended his central register to make use of its information in a statistical way:

The numbers of people that are of every yeare old from one

> to 100, and the number of them that dye at every such year's age, do shew to how many year's value the life of any person of any age is equivalent and consequently makes a Par between the value of Estates for life and for years' (*The Petty Papers,* vol. 1, p. 193 Cambridge UP, Cambridge 1899).

Petty was exclusively interested in the making of inferences from any data that was collected by the method of asking questions of it. When someone begins to search data for non-obvious answers to interesting questions, it is a clear indicator that a statistical approach to numbers has replaced that of merely glancing at the bottom line total.

The techniques used for searching data do not need to be sophisticated for them to be regarded as evidence of a statistical approach. Considerable progress can be made, even by the simplest of manipulative techniques. For instance, Petty proposed to take proportions of mothers to births to see what effect abortions and 'long-suckling' had on the 'speedier propagation of mankind'.

Comparing samples

He also proposed to subject his own profession, that of the physicians, to what amounted to a crude efficiency test (how popular such a test would have made him among his colleagues we can only guess at, but it was a brave example to offer). As Petty's actual medical qualifications in this area leave much to be desired – at best they can only be described as of dubious relevance – his interest may have been motivated by knowing how ill-equipped he and his contemporaries really were to treat the sick!

One of the tests he proposed was to find out:

> Whether of 1000 patients to the best physicians, aged of any decade, there do not die as many as out of the inhabitants of places where there dwell no physicians.

If there was a large difference in the rate of survival between the two circumstances of having and not having a physician's services, this might confirm whether being attended by a physician was a sensible investment or its reverse. If there was no significant difference in the survival rate, it might suggest that the sick would be better off holding onto the fees they would normally pay to the medical profession for services which gave them no greater chances of survival than if they had let nature take its course.

We can see here the embryo of the statistical approach to the comparison of two populations, one of which has an identifiably different characteristic from the other – in this case, the presence or absence of medical attention. This method can easily be generalized for almost any other comparative characteristic, such as age, gender, family background and so on. Nowadays it is commonplace to make comparisons in this way, though not always with a strict partiality for what is appropriate. We might want to know whether graduates perform better in management than non-graduates, or whether graduates have higher lifetime earnings than others. Does the use of a drug, for instance, improve the chances of recovery? This could in principle be tested by recording the medical effects of a group of patients who take the drug with another group who do not. This method of the use of controlled sample is widespread today. Until Petty the concept had not been formulated, let alone applied, and his method, therefore, is much more important than his proposed application.

Petty's overall contribution to the emergence of statistics was to make radical proposals rather than achieve practical results. In fact, much of his practical statistical work borders on (and occasionally crosses into) fantasy.

National income analysis

Petty did make, however, an important contribution in the realm of national income analysis which illustrates another early application of the statistical approach and one that was to be more fully developed much later, particularly after the Second World War.

Whereas Graunt took current and available data and made statistical inferences from it, and in so doing took the first step to statistics, Petty, on the other hand, identified some results that would be of interest if they were available, and with this objective in mind he proposed that the necessary data be collected. His recognition of the need for a competent and authoritative data source was the second step towards statistics.

Petty called this type of activity *political arithmetic,* and, as I have noted, some regard this as the beginning of economics. His method was Baconian in tone. Indeed, Petty noted Bacon's 'judicious parallel between the Body Natural and the Body Politick'. He went on:

> Now, because Anatomy is not only necessary in Physicians, but laudable in every Philosophical person whatsoever; I therefore, who profess no politics, have for my curiosity, at large attempted the first Essay of Political Anatomy' (preface, *The Political Anatomy of Ireland,* (1670) 1691).

Petty's main contribution to political economy was to attempt to prove certain propositions, all of them conducive to the good government of Charles II and also critical of the *impressions* that appeared to be widespread about the effects on Britain of the prolonged constitutional troubles it had experienced. For the moment we can ignore an inclination to react sceptically to his trying to prove something he believes beforehand; we shall consider the consequences of his approach rather than concern ourselves with his motives.

Petty did not want to merely make unsubstantiated assertions (though he does plenty of that) nor to argue speciously for propositions that could not be tested empirically. He set out to establish his points by appealing to the evidence, or the best interepretation of the evidence that was available. His was a statistical argument – based on inferences from the data – and it is this that makes Petty's writings on national income so important for statistics.

His works that are relevant here are his *Verbum Sapienti* (1665) (Wise Words), and *Political Arithmetick* (1676). In contrast to the prejudices that eventually arose about economics, Petty set out to confute 'dismal suggestions' which 'I find too currant in the World, and too much to have affected the

Minds of some, to the prejudice of all'. His is a plea for optimism, quite contrary to later images of economists as exponents of the 'dismal science'.

In *Political Arithmetick,* Petty puts forward ten 'Suppositions', that allegedly refute the pessimism about England's economic fortunes and its relative economic and military strengths against that of France and Holland. No wonder the King of France was 'offended' by privately circulated versions of Petty's conclusions and, in consequence, ensured through diplomatic displeasure the delay in its publication until 1690.

To establish his main data base, upon which the others stood or fell, Petty had to make an estimate of England's wealth. But there were no national income statistics conceived of, let alone available, in the seventeenth century, and he had to start from scratch.

His first assumption was in itself revolutionary. He assumed that the national income was the same as its consumption and that the amount of saving out of incomes could be disregarded safely given the troubled years the country had lived through. From this it was breathtakingly simple: how much per head did the people spend and how many people were there?

He had no real figures so he estimated them at £6.13s.4d per annum for every man, woman and child and multiplied that by an estimated 6 millions of people in England and Wales, giving an annual expense of £40 million (an estimate he increased to £42 million in 1676).

He was convinced that 'the Power and Wealth of England, hath increased above this forty years' and that 'there are spare Hands enough among the King of England's Subjects to earn two Millions per annum more than they do now, and there are Employments, ready, proper, and sufficient, for that purpose'.

As for income, he estimated this to consist of £25 million as the wages of labour and £15 million as the income from property. He supposed that if all employed people were taxed at 10 per cent of their earnings, this would raise £4 million for the state, which was 'sufficient to maintain one hundred thousand Foot, thirty thousand Horse, and forty thousand Men at Sea, and to defray all other Charges, of the Government: both Ordinary and Extraordinary' (ibid, 1676).

Petty distinguished between income and what he called stock, or property. The Kingdom's property yielded £15 million a year and was worth in total about £250 million by his estimates (which he got by estimating the annual rents of all the agricultural land, the rents of all the houses, and the income from cattle, commerce and trade). This gives a rate of return of 6 per cent (that is 15/250 = 6 per cent).

Next, he asked what stock would be required to earn the £25 million that went to labour if the rate of return is 6 per cent? This gave him a notional value of the stock of labour of about £416 million. The calculation is ingenious, and, as a direct application of his method of questioning the data, it is another illustration of the difference between numbering and statistics. Consideration of the economic implications of the calculation would take us too far away from our theme, though economists might want to ponder it.

Interestingly, from the point of view of deriving conclusions from the data, he also estimated that the money loss to the country of the 100,000 who died of the plague was worth about £7 million (that is, their annual incomes times their average expectancy of life – which was about 8 years in Graunt's tables for the working ages of 16 to 56). In common with his reforming spirit, Petty suggested that a modest expenditure of £70,000 by the state to prevent deaths from the plague would be of immense economic benefit greatly in excess of its cost.

Gregory King's statistics

Some of the ideas held about statistics, even by those well versed in them, were quite fantastic. We have already noted Arbuthnott's belief that a statistical regularity (the excess of births of boys above that of girls) represented evidence of a divine will at work in the universe. He was not alone.

With the progress of statistics in the field of probability, belief in an inherent order that produced natural 'laws' behind the seeming chaos (though not necessarily being seen as something divine in origin) became a source of great inspiration among statisticians to derive mathematical explanations for what they were finding by observation.

Petty's work on national income statistics was followed within six years by an even more impressive estimate of England's national income by Gregory King (*Natural and Politicall Observations and Conclusions upon the State and Condition of England* (1696), published in 1802). At first it circulated privately and in extracts published by Charles Davenant (1656 – 1714), and it stuck very closely to the data without the flights of fancy typical of Petty's work.

King had strong views on the efficacy of national income estimates for the state and it is most likely that similar considerations persuaded other states to try similar exercises. King writes:

> If it be well apprized of the true State and Condition of a Nation, Especially in the Two maine Articles of it's People, and Wealth, be a Piece of Politicall Knowledge, of all others, and at all times, the most usefull, and Necessary; Then surely at a Time when a long and very Expensive Warr against a Potent Monarch ... Seems to be at it's Crisis, Such a Knowledge of our own Nation must be of the Highest concern (King, preface, 1696).

He also justified the method of estimating data that was not otherwise available because

> We must Content our selves with such near approaches to it as the Grounds We have to go upon will enable us to make.

King's is a far more elaborate estimate of the national income than Petty's, supporting the view that once a scientific method is developed it will be improved upon. He used a lot of available data – the polltax accounts, the hearth, birth, marriage and burials taxes, the excise taxes on consumables and customs duties, estimates of the value of various national outputs (mainly agricultural), and estimates of national expenditures. With much careful work he estimated the national income for 1688 at £43,505,800 over a population of 5,500,520. The annual income per head was £7.18s and the annual expenses were £7.11s.3d (King, *Two Tracts,* 1936).

Samples of one

Charles Davenant defined the use of political arithmetic as 'the art of reasoning by figures, upon things relating to government' (1698). In other words, political arithmetic was an aid to decision-making. He also outlined what was involved in the techniques of political arithmetic, and in doing so, set out for the first time the methodology of statistical operations:

> He who will pretend to compute, must draw his conclusions from many premises; he must not argue from single instances, but from a thorough view of many particulars; and that body of political arithmetic, which is to frame schemes reduceable to practice, must be composed of a great variety of members. ... A contemplation of one object, shall give him light into things perhaps quite of a different nature: for as in common arithmetic, one operation proves another, so in this art, variety of speculations are helpful and confirming to each other (Davenant, 1698, vol. 1, pp.145 – 6).

The number of times that Davenant's strictures against arguing from a sample of one are stated in an elementary statistics course is, in my view, a measure of its value.

Exercise 3.1

The most common type of single-sample argument in my experience goes something like: 'my granny smoked forty a day and died aged 96, so smoking cannot be bad for your health'. Perhaps you may care to pause here and list the reasons why such a statement has little validity? (I give mine in the Appendix).

The spread of the census

The seventeenth century saw the emergence of state census activities in Europe. Denmark tried a census in 1645 to establish a taxation base (repeating it in 1660), and Norway surveyed the number of males eligible for military service in 1662. Iceland held a census in 1703 but did not apparently use it (probably from a lack of resources to do so) and in Sweden in 1748 the government decreed that population records were to be kept.

A bill to hold a census in Britain in 1753 failed to get support in the House of Lords, while Arthur Young's many pleas for a census fell on deaf ears, though the canny Scots managed to use the services of their formidable church ministries to make a rough and unofficial estimate of the Scottish population in 1755. The lack of evidence in British population statistics was no barrier to debate and learned assertions, and it is against this background that Thomas Malthus wrote his famous *Essay on Population* (1798), which sent off a large number of hares round the world, some of them still running.

The United States included in its constitution provision for a census (Article 1, section 2). This required: 'that the actual enumeration shall be made within three years after the first meeting of the Congress of the United States, and within every subsequent term of ten years, in such manner as they shall by law direct'. The first census took place in 1790 and has been faithfully followed every ten years since.

Work on estimating mortality rates continued. A Frenchman, Deparcieux, found an ingenious method of estimating a true life table. He studied the mortality of a Benedictine monastery for the years 1607–1745. This sample gave him an isolated and stable community – mobility of the people in a sample is always a problem in an open city – and he followed it up with other studies of *tontines* (those lotteries where the subscribers agree that the survivors of an age cohort are the beneficiaries, that is, the last one alive scoops the jackpot, even if too old to enjoy it!).

In the Netherlands, Struyck (1687–1769), Kersseboom (1691 –1771) and Van der Burch (1673–1758) published several items on mortality statistics, largely developing Graunt's work, and in Britain, Halley, among others, published a life table on the basis of data sent to him by a clergyman in a town in Germany and he also corrected the errors in Graunt's.

In Germany, J. P. Sussmilch (1707–67) made a major contribution to the burgeoning science of statistics. Like others he was obsessed with the evidence of a 'divine order' (the title of his book (1761)) applying to population, mortality and diseases. He was also adamant that polygamy was obnoxious to God's will – why the seventeeth and eighteenth centuries were so uptight about polygamy could be an interesting study for a sociologist, or, more likely, a social psychologist, though it is a pity we will probably never know the private views of eighteenth century women on the subject.

The lady with the data

Those of you fortunate to have access to a Bank of England £10 note will find on the reverse side a portrait of one of Britain's most effective statisticians of the nineteenth century, Florence Nightingale (1820–1910). You may not think of her as a statistician, and the designers of the £10 note do nothing to enlighten you about this other role in her life, for they show her, complete with the inevitable lamp, in a military hospital tending the sick.

Florence Nightingale realized the power of statistics to support her long campaigns for the total reform of the hospital system. For her, credible and accurate numbers were ammunition to fire off at the silly asses who ran a hospital system that was more lethal to the patients than the worst efforts of a determined enemy to kill and maim them. Why credit disease with being the cause of death when it was the administrators who were culpable?

During the first seven months of the Crimean War, the mortality rate from disease alone reached an unbelievable 60 per cent. She also deduced that if the army treated itself this way in the field then it was likely that it was not much cleverer back at barracks, and sure enough, once she had checked through the figures, she found a mortality rate of soldiers between 25 and 35 years old that was twice what would be expected among a comparable group of civilians.

You can sample the concentrated anger that typified her campaigns to get the top brass to see reason in a letter she wrote to Sir John McNeil:

> it is as criminal to have a mortality of 17, 19 and 20 per thousand in the Line, Artillery and Guards, when that in civil life is only 11 per thousand, as it would be to take 1,100 men out upon Salisbury Plain and shoot them.

The state of Chatham Hospital (laughingly regarded as a refuge for sick soldiers) spurred her to remark:

> This disgraceful state of our Chatham Hospitals is only one more symptom in a system, which in the Crimea, put to death 16,000 men – the finest experiment modern history has seen upon large scale, viz., as to what given number may be put to death at will by the sole agency of bad food and bad air.

Nightingale's campaigns for hygiene in hospitals and barracks exposed the deficiencies of the statistics as much as those of the wards and billets. It is not possible to have statistics without data and the data has to be accurate if the inferences that are obtained from it are to be valid. Nightingale's efforts to get reliable data illustrate a problem that faces anybody who wants to work over data for some statistical purpose.

Statistical records have to be uniform. It is no good having one set of figures using one system of classification and other sets using other systems. The data must be compatible, otherwise it is impossible to conduct even elementary operations like adding and subtracting.

Nightingale wanted each hospital to use the same system of classification and the same nomenclature. To this end she designed a standard form for hospitals to use. One of her suggestions was for an entry to record the diseases that patients contracted while in hospital in addition to those that they brought with them, in an effort to highlight how successful the hospitals were at causing gangrene and septicaemia!

Statistics is not just about enumeration of facts. Nightingale recognized the difference between enumeration and statistics:

> The War Office has some of the finest statistics in the world. What comes of them? Little or nothing. Why? Because the Heads do not know how to make anything of them. Our Indian statistics are really better than those of England. Of these no use is made in administration. What we want is not so much (or at least not at present) an accumulation of facts, as to teach men who are to govern the country the use of statistical facts.

That Nightingale understood how to use statistical facts is evident from the kind of questions she thought should be tackled if the data could be collected.

In the proposals for the 1861 Census she asked for questions to elucidate the number of sick and infirm in the land and also for details of housing. What she wanted to test was a hypothesis about the connection between health and housing conditions. This is how a statistical mind operates, looking for connections between things and searching out ways of getting accurate data. Unfortunately, the House of Lords lived up to its nineteenth century reputation as a bastion of progress (as long as nothing changes) and put the kibosh on her proposals.

Undaunted, Nightingale pressed for other questions to be tackled by the statistical method:

What had been the result of twenty years of compulsory education?

What is the effect of townlife on offspring in number and in health?

What are the contributions of the several social classes to the population of the next generation?

Again, we can see the statistical mind at work: formulation of a hypothesis - search for data - measurement operations on the data - derivation of the conclusions. (Perhaps I should add that a statistical approach also requires a willingness to accept the conclusions, even if they cause discomfort in respect of our preconceived notions, though I recognize this sometimes can be worse than having toothache.)

A graph is worth a thousand numbers

The use of graphs and diagrams to show the salient features of data is well established in our daily lives. Hardly a day or two goes by in which we do not see a graph or diagram in a newspaper, magazine or on television. It has not always been thus. Visual presentation of data did not become common until late in the nineteenth century. The first fifty volumes of the *Journal of the Royal Statistical Society* (1837–87) only contain about fourteen charts, with the first one appearing in the 1841 volume.

Nowadays, no course in statistics is complete – it can hardly be said to have begun – without some work on the construction of what are histograms, frequency curves and, the delightfully named, ogives.

Honour for first place in the Pantheon of statistical graphologists is a close tie between a German, A. F. W. Croome (1753–1833), and a Scotsman, William Playfair (1759–1823). Croome appears (technically at least) to have just pipped Playfair at the post by publishing in 1785 a chart (Grossen Karte) illustrating the comparative sizes of European states. Croome explains that he chose to use his chart on the grounds that

> the different sizes can however be more easily seen and grasped if they are brought before the eye in the form of a drawing, because the imagination is thus stimulated, than if these merely appeared in the form of numbers, especially when these consist of many digits as is often the case with the areas of states (1785, quoted in Royston, 1956).

A similar motivation appears to have inspired Playfair to make use of his charts. He wanted them to make statistics more palatable to the consumer:

> for no study is less alluring or more dry and tedious than statistics, unless the mind and imagination are set to work or that the person studying is particularly interested in the subject; which is seldom the case with young men in any walk in life (Playfair, 1801).

William Playfair appears to have been a trifle off-white-going-on-grey sheep of his family. Unlike his distinguished elder brother, John Playfair, Professor of Natural Philosophy at the University of Edinburgh, William was under a cloud of doubt during most of his eratic life. That he was bright, entrepreneurial and energetic is clear from his activities. He was also opinionated, politically inclined (he lived and politicized in revolutionary France for some years until he fell out with the revolutionaries), and he was prolific as an author.

He fell out with the Edinburgh literary establishment after publishing some derogatory remarks about Adam Smith and he was not taken all that seriously elsewhere as a result. This is a pity because he invented the use of time-series charts in presenting statistical data (whereas Croome may be said only to have invented the *idea* of graphical illustrations – his original system of 1785 did not catch on).

Playfair first used charts in his *Commercial and Political Atlas* (1786). This volume contained 44 charts, one of which was the first ever use of a bar chart. He called his method 'lineal arithmetic' and you can get an idea of what is the thinking behind the construction of a time-series bar chart from his own description:

> The advantage proposed, by this method, is not that of giving a more accurate statement than by figures, but it is to give a simple and permanent idea of the gradual progress and comparative amounts, at different periods, by presenting to the eye a figure, the proportions of which correspond with the amount of the sums intended to be expressed.
> Suppose the money received by a man in trade were all in guineas, and that every evening he made a single pile of all the guineas received during the day, each pile would represent a day, and its height would be proportioned to the receipts of that day; so that by this plain operation, time, proportion, and amount, would be all physically combined. Lineal arithmetic then, it may be avered, is nothing more than those piles of guineas represented on paper, and on a small scale, in which an inch (suppose) represents the thickness of five million of guineas ... as much information may be obtained in five minutes as would require whole days to imprint on the memory ... by a table of figures' (quoted in Fitzpatrick, 1960) (see figure 3.1).

Playfair, in various editions of his works, introduced into statistics the line graph, circle graph (which Croome appears to have copied in 1820), bar chart, and the pie diagram, most of which are now common fare in statistics.

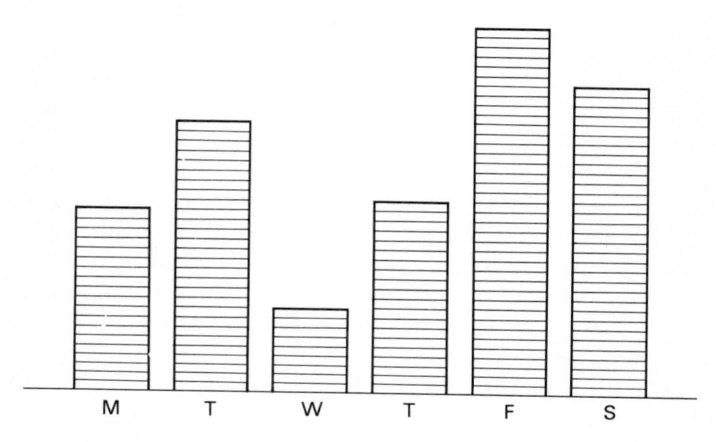

Figure 3.1 Playfair's trader and his weekly takings

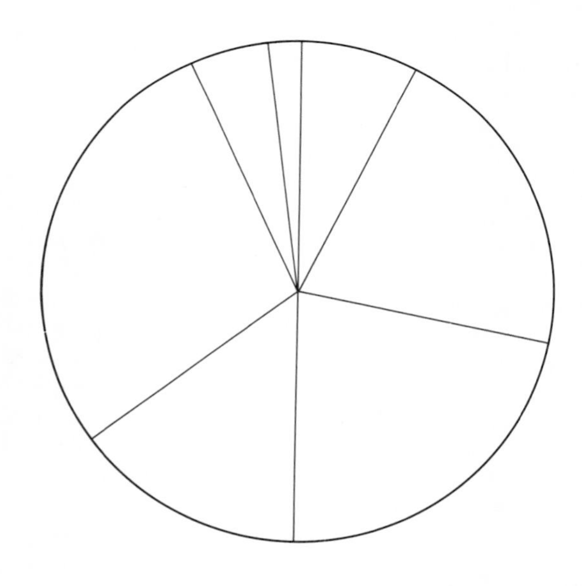

Figure 3.2 Pie diagram – each segment size proportional to share of total

The pie diagram is probably the most common device used in simple statistics. (figure 3.2).

The principle behind a pie diagram is to divide a circle (pie) into segments that are proportional to the whole. Thus, if half a company's sales were in detergents, then the segment marked detergents would form half of the pie, if it was a quarter it would be a quarter of the pie, and three-tenths, three-tenths of the pie, and so on. It is the most simple of constructions and does strikingly reveal the relative proportions at a glance.

Exercise 3.2

Try a little graphical exercise: suppose that you were the marketing director of a chemicals company and you wanted to show your fellow directors how annual company sales were divided between the various products. If plastics accounted for 35 per cent of the sales, oil cake 25 per cent and processed oils 40 per cent, how could you represent this breakdown using a pie diagram? See the Appendix when you have completed your own diagram.

On Playfair's own admission he got the idea of charting from his elder brother John, who, taking over the job of bringing up the family when their father died, had made 13-year-old William keep a register of the readings of the house thermometer to help educate him (or more likely, to find something to do for his mischievous idle hands). John told him to represent the temperature variation by lines drawn to scale and averred that 'whatever can be expressed in numbers may be represented by lines' (1805, quoted in Royston 1960). There is also a likelihood that during his time with James Watt as a draughtsman, William learned about the use of graphs to represent the relationship of steam pressure to volume.

Only Playfair's known activities in other areas – where he was a trifle short of probity in respect of other people's works – has precluded an open-arms acceptance that he was the inventor of graphical methods in statistics. What is beyond doubt, however, is the fact that Playfair was the first to publish the types of graphs that are common in statistics today.

Historical note: who invented statistics?

There has been much debate on the origins of statistics. There are several strands to it and I think it useful to round off this chapter with a brief summary of the main issues because it is yet another illustration of the difference between statistics and mere enumeration.

The word statistics has a Latin root in the word *status* (state) and *statista* (statesman). The Greeks were interested in comparing one state of society with another (Aristotle – the originator of the deductive syllogism – described 158 different types of states). From this tradition of comparative constitutions, a German school of statistics developed about the same time as the statistical work of John Graunt.

The German school was influenced by Hermann Conring (1606–1682) who was a professor at the Brunswick University of Helmstaedt (Lazarsfeld, 1961). In 1660, he began lecturing on political constitutions of states and notes from these lectures were circulating among students in Germany for many years afterwards. Such was Conring's influence that his ideas were taught elsewhere in Germany.

A pupil of one his students, Gottfried Achenwall (1719–1772) published a book in 1749 that was about 'The Science of Today's Main European Realms and Republics'. Many authorities have credited Achenwall with the first use of the word statistics and accordingly have declared him to be the father of the subject. Unfortunately, it is clear that Conring preceded Achenwall in the subject matter of his life's work and that the subject matter itself makes a nonsense of the claim that the coining of the word statistics by Achenwall had anything to do with the subject of statistics as we know it today.

We can go further, but not in detail as we have much else to consider, and state categorically that the German school's pretensions to the discovery of statistics falls on two simple grounds.

First, the Germans who contributed the name (but not the subject) of statistics had it filched from them by a Scotsman, Sir John Sinclair, who literally lifted the name which they applied to discussions about states (and why not, they had a country that consisted of 300 separate states and they were obsessed by them!) and applied it to a project of statistics proper, namely his *Statistical Account of Scotland* (1791–99). He did this without blushing and was always very proud of his 'theft', much to the annoyance of the Germans. He was looking for a catchy name for the series on the new subject he was about to publish, and he was certainly right in his choice, for *statistics* caught on and became identified totally with the inferential work begun by John Graunt.

Secondly, in reaction to this 'misappropriation' of their title for their exhaustive studies of states, the Germans hit back in outbursts against the statisticians proper which fully exposed their lack of claim to anything other than the title. For example, they wrote of the practioners of numerical statistics:

> These stupid fellows disseminate the insane idea that one can understand the power of a state if one just knows its size, its population, its national income, and the number of dumb beasts grazing around' (quoted in Lazarsfeld, 1961)

This was by no means the least of their venom:

> The machinations of those criminal politician-statisticians in trying to tell everything by figures ... is despicable and ridiculous beyond words (ibid.)

The evidence is therefore conclusive.

Statistics as a subject that derives meaningful conclusions from numerical data was developed first by John Graunt and later spread around the world with his approach as its agreed content. It achieved the title of statistics in 1791 when Sir John Sinclair lifted it from German university studies of the nature and consequences of states. The Germans reacted bitterly to this, simultaneously claiming to have originated the title statistics (correct) and the modern subject matter of statistics (incorrect) and at the same time denoucing the modern numerical methods in terms that leave no room for misunderstanding.

4

The Birth of Probability

It is a remarkable coincidence that at precisely the same time as John Graunt was compiling his manuscript on the Bills of Mortality, across the Channel, in France and Holland, various individuals were working on century old problems of games of chance, and their efforts led eventually to the theory of probability. This is a remarkable coincidence and typical in itself of what probability is all about, and though drawing inferences from data and the solving of brain teasers about chance were entirely separate in conception and content, they were destined to fuse together into the subject of statistics.

What is the connection between the manipulation of data and the theory of probability? In large measure the connection arises because of the nature of real world data. Circumstances often conspire to prevent access to all the relevant data that a statistician might be interested in and he or she has to make do with a large or small sample from a much larger *population*. In John Graunt's case he made estimates of the total London population by taking samples of it. Alternatively, in theory, he could have counted every single person in London and arrived at the total population without using samples. Realistically though, there was no way that Graunt could have counted everybody, and much the same problem is faced by every scientist who studies any large population (and note, *en passant,* that in statistics any quantity of things, be they people or cars, or atoms, or seeds and so on, is called a *population*). In the absence of a practical means of getting at the total population we are forced to use samples. Moreover, and this is the relevance of the study of probability, it is not necessary to study a whole population even if we could do so, because by using probability theory we can make relatively reliable predictions about the characteristics of a population from samples from it. There is a trade off between certainty and likelihood when we use a sample (from circumstance or choice) but such is the relative reliability of the probability calculus when applied to statistics, that we can choose the degree of confidence we require in order to make statements about a population's characteristics on the basis of a sample from it.

The fusion of probability with data had a tremendous impact on statistical techniques and for this reason alone a study of the origins of probability theory will pay large dividends when you move on to examine specific probabilitistic techniques. Our story opens in an unusual setting, namely the study of games of chance, and it discusses the plays of dice and cards. For a while this may seem to be unconnected with the science of statistics, but the discoveries over the centuries about the behaviour of chance devices (such as dice and cards) and the magnificent leap from real devices to their abstract theoretical counterparts (the imaginary die and the imaginary pack of cards) replicates vaguely that leap centuries earlier between arithemtical numbers and a universal geometry. What theory did for mathematics, abstract theories of chance did for statistics. That the latter had its origins in the relatively trivial world of gambling is a footnote in the history of statistics, but an interesting one at that.

'The decade around 1660 is the birthtime of probability' (Hacking, 1975, p. 11) and the question that must be posed, because it is of more than curious importance, is why did its birth occur at this time and not previously?

In this and the next chapter, we shall explore the work of the continental mathematicians (do not panic, I do not intend to impose upon you their mathematics) and how they were led to the probability calculus, or, more exactly, the science of the *definite maybe* (apologies to Sam Goldwyn).

The long pregnancy of probability

'Games' of chance have been around for more than three

thousand years according to David (1955, 1962), citing archeological evidence from Pharaonic Egypt. I have put games in quotation marks only to draw your attention to the real possibility that for many of the people concerned the act of using chance mechanisms may have had much more to do with religion or superstition than with games as we might think of them nowadays, that is, entertaining ways of amusing ourselves by 'winning' or 'losing' something notional (points) or real (money). It is probably more correct to refer instead to *devices* of chance, but this might, on the other hand, be a trifle pedantic and can be avoided if we bear in mind what we mean by games.

The Greeks played various games with astralagi (bones from the heel of a sheep or dog, popularly known as 'knucklebones'), the sides of which were identifiable, though not uniform, and which were given various values depending upon the combination that they fell in. The worst combination of four thrown astralagi was known as the 'dog' or 'vulture', and the best of all was 'Venus', when all four uppermost sides of the bones were different.

Pottery dice, complete with dots on the six sides from one to six have been found in Iraq and India and have been dated as third millenium, and there are throwing sticks among the relics of ancient Britons, Egyptians, Romans, Greeks, and Maya Indians in America.

The Bible reports on the Israeli trust in the 'casting of lots' to make important decisions, such as, for example, the division of the land of Israel between the tribes and families (Numbers, 26, v. 55: 'the land shall be divided by lot'), and the settlement of disputes (Proverbs, 18, v. 18: 'The lot causeth contentions to cease, And parteth between the mighty').

In Roman times the passion for gambling was in full flood. Augustus wrote to Tiberius about his gambling:

> We gambled like old men during the meal both yesterday and today, for when the dice were thrown whoever turned up the 'dog', put a denarius in the pool for each one of the dice, and the whole was taken by anyone who threw the Venus' (quoted in David, 1955).

Claudius wrote a book about dicing and played while driving

his carriage, and some travelling Romans took their croupiers along to roll dice for them. Hence, the evidence is clear that devices for games of chance in some form or other were used in very ancient times for either amusement or spiritual advice.

During the dark ages, the church authorities frowned upon gambling, as did the Roman authorities, and so have most guardians of morals since. Why the powers that be frown so uniformly on a natural pursuit of people when left alone, escapes me, particularly as history shows that the authorities themselves are not immune to gambling privately (viz. Augustus and Claudius above), and that by making it a crime they create a lucrative living for criminals, and make honest people into 'criminals' by silly laws, which is presumably not what they intend. Most of the disagreeable side-effects of gambling's 'vices' are caused, in my opinion, by the prohibition itself, not by the activity.

St Cyprian of Carthage (*circa* AD 240) attacked the vices associated with gambling, as did St Bernadino of Siena in 1423 in similar terms, indicating that gambling must have been prevalent for the vices to have been practised. Crusading knights and clergy were restricted by laws in their gambling (typically the common herd were banned totally from the practice) to losses of no more than twenty shillings a day (1190). King Fredrich III issued a law against it in 1232, as did Louis IX in 1255 and edicts were passed against the clergy indulging in such games in 952, 1227, 1238, and 1240 (Kendall, 1956). Edward III prohibited gaming, and Henry VIII declared dicing and cards 'unlawful' for his subjects, though typically not for himself.

As a 17-year-old in Australia, a high point in my adolescence was to participate in illegal 'two-up' games in the basement of a restaurant behind Sydney's Central Railway Station. In 'two-up', two coins are tossed in the air and the people make bets on them falling 'odds or evens'. The 'house' takes a percentage of the stakes but otherwise the individuals settle with each other. If you won a lot of money – and I once saw a man win £2,000 (in 1958 prices) – the house guaranteed you a safe passage home, escorted by a 'gorilla' in a taxi, with nobody else permitted to leave the room for 30 minutes. It was widely believed (though I personally never tested it) that once a year on Anzac Day the police would not arrest people playing two-up in public, though I must confess that playing it legally would have taken away half the fun for me, and, because I had first-hand experience of *professional* gaming people, I would have been less sure of the honesty of the game!

Undoubtedly, gaming was a common pastime in Europe among gentlemen and philosophers and not merely a bad habit of thieves and vagabonds. However, the existence for many centuries of a mechanism for deriving chance events, and its fairly widespread use for good or ill among all layers of the population, did not produce a mathematics of chance. This is of great significance to the study of the development of probability.

Consider the case of Darwin's theory of natural selection to explain evolution over millions of years. In the absence of noticeable change in species within the lifetimes of observers, we can perhaps account for the delay in discovering evolution until the nineteenth century, when a comparative observation across similar species in widely separate environments was undertaken, but in the case of games of chance, large numbers of people repeatedly observed the outcomes of millions of games and, in addition, they had a financial interest in correctly predicting the outcome, and yet the simple arithmetic of chance did not appear in a lasting (that is, correct) sense until around 1660.

Observation, as we have noted, is not enough; the observer must also be looking for something. Vague notions about chance did exist – what is a bet between players on an outcome if it is not a notion of chance with each player believing he knows the forthcoming outcome more accurately than the other? But a theory of chance did not exist, nor did there appear to be any pressure to discover one. Gambling by itself did not and could not produce a theory of probability, though it could and did become the vehicle through which the theory was launched.

The recognition of cases

The first steps towards a theory of chance required that the possibilities of chance events be recognized, and, for ease of study and comparison, that they be listed by somebody. This was not so obvious then as it might seem to us looking backwards with the benefit of hindsight. It certainly was not obvious to the gamblers, though they developed, through empirical evidence of the games, notions about relative probabilities to the extent that they awarded greater pay-offs to some throws of dice, or named hands of cards, than the others.

We know that the empirical 'rules' of the gamblers were not theories of probability, because for a century or more mathematicians were required to demonstrate their prowess by explaining them theoretically.

In the quest to solve certain well-known problems of chance in relation to gambling, some people have assumed that the theoreticians themselves were either practicing gamblers or were responding to problems set by practising gamblers. In this I think the point is being missed – if gamblers wanted these problems solved why did it take centuries to do so and what did they do in the meantime? In my view these problems were only tenuously related to real world gambling and were in the main what we would regard today as 'brain-teasers'.

The differences in the odds between some of the outcomes posed by the problems are so slight, and might require such an enormous run of games for them to be of financial benefit to the throwers, that it is unlikely that practising gamblers would have the patience to play such games to their conclusion to win what could amount to derisory prizes.

My inclination, based on my own experience (and having no reason to believe that fifteenth-century gamblers were any different in outlook), is to believe that players are only attracted to games that are short enough to give frequent pay-offs and which give them a chance (however remote) of really big prizes.

Of course, in the case of the gambling 'house' the motivation is somewhat different. A house makes a *certain* profit by taking small percentages on *every* throw rather than uncertain big percentages by risking its own money. Houses provide the opportunities for people who are fired with the fantasy of the 'big score'; they do not themselves become victims of their own fantasies and they avoid doing anything so rash as to gamble themselves – at least not those houses that want to remain in business and grow rich quietly. In roulette for instance, the house takes all the stake money if zero comes up, thus increasing the odds against a player having a successful bet from 1 in 36 to 1 in 37. In other words, gambling houses are not run by gamblers, or, pehaps more accurately, they are not run by gamblers for very long!

The kind of problems that mathematicians of the sixteenth and seventeenth centuries attempted to solve were theoretical rather than practical; they used games of chance as the instrument of the chance experiment rather than as the object.

What then was the importance of games of chance to probability? First, regularity of outcomes had to be noticed and then recorded, and games of chance are an ideal experimental device for identifying such regularities. It is interesting (and almost inevitable) that the first appearance of regularities had applications in the context of ascribing fortunes or consequent actions to the individually different outcomes that were possible in the throws of dice, or the hands of cards. This habit of linking fortune to the toss of a die, or the fall of a card, has survived throughout the centuries and can still be found in the fortune teller's tent at your local fair, or in occult corruptions of the noble and ancient game of tarot (first played in northern Italy *circa* 1430).

Progress has many instruments and if fortune telling was one of them for a while it is no cause for dismay, because in listing the possible outcomes of a game of chance, the fortune teller unintentionally made known the fact that the outcomes in a game of chance could be counted.

Cases

I shall call the outcomes of the throws of dice cases. Thus, for one die there are six cases, as only one of six faces can be uppermost per throw. In the throwing of three dice, a 3 has only one case, namely, 1, 1, 1, because it cannot be made any other way. Similarly, though a 4 can be made with only one set (or partition as the jargon has it), namely, 1, 1, 2, it can be made in three different ways: 2, 1, 1; 1, 1, 2 and 1, 2, 1. We say, therefore, that a 4 has three cases.

Three dice can land uppermost with, say, 1, 2, 3 showing, or with 4, 5, 1, or 6, 6, 1, and so on. Each one of these throws can be obtained from different combinations, which are different ways in which a number can be made by throwing the three dice. The more dice, or the more cards, the greater the number of possible cases. For two dice there are 36 cases, and for three dice there are 216. How these were discovered might interest you.

Kendall reports evidence that many of the cases of the throws of dice were known throughout the Middle Ages. Cases are important (and feature in elementary probability exercises) because it is essential to be able to enumerate all the possible throws of the die before an idea of the frequency of particular throws can develop into a probability calculus.

Evidence of the knowledge of the existence of cases is found in a manuscript that lists the '56 virtues', one for each of the ways that three dice were believed to be thrown. Apparently, some monks used it to decide which virtue they would practise for the day by throwing three dice and then consulting the list for the appropriate case against which there was a prescribed virtue.

The falls of the dice are considered irrespective of their order and this led to many possible cases being ignored by the monks. Thus, a throw that produced 1, 2, 3, was counted the same throw as one that produced 2, 3, 1. For a full list of all the possible cases the order of the dice would have to be considered, because each die is a separate chance device, though it would be difficult to separate them out without some distinguishing features.

If this point is somewhat obscure, consider that you are throwing three dice, each distinguished by a different colour. As the dice fall you list them in columns and always enter the number of the blue die in the first column, the red in the second and the green in the third. Clearly, throws of 1, 2, 3 and of 3, 2, 1 under this system are different cases and must be listed as such.

It might be more obvious that this is a sensible distinction to make if you think of the three windows through which you view the spins of the lemons, plums and bars in a slot machine. If instead of lemons, plums and bars you had numbers, it ought to be obvious that if the spin settled with an order like 1, 2, 3, this would be a different order from one like 2, 3, 1 because this spin results in the same numbers in different windows.

That the order of throws would not be conceived as being important initially, is not surprising – it is a slightly more abstract way of looking at the throws, and one that gamblers (or monks) may not consider important as they throw indistinguishable real dice and not distinguishable abstract ones. In the actual throwing of dice it is not obvious which die is which compared to a previous throw, as they are mixed up when shaken by the thrower (assuming true and not trickster dice; see Scarne (1962)).

It is also possible that the author(s) of the religious manuscript had enough trouble devising 56 virtues without having to stretch even their well-known patience by thinking up as many as 216 of them, and that they were (understandably) inclined to limit their list of throws to the unordered ones!

Exercise 4.1

If you want to get an idea of what is involved in listing the cases of a chance device and you find yourself with nothing to do for a while on a train, or in a lecture, why not write down, as a doodling exercise, as many ways as you can in which three dice can land, ignoring for the moment the order (that is, count the throw 1, 2, 3 the same as the throw 3, 2, 1 or 1, 3, 2). If you are at least as good as the uncouth monks of the fifteenth century you should get 56 individual cases, eventually (see Appendix for my listing).

St Bardinino, mentioned above, referred to 21 (unordered) throws of two dice (there are actually 36 ordered cases) in the course of his sermon against the evils of gaming, indicating that in his ignorance he knew something very important.

Exercise 4.2

You could, if you prefer, have a go at the relatively easy 21 cases of two dice, and then try for the full 36 (see listing in Appendix).

Exercise 4.3

If you return to the three dice, but imagine that each one is of a different colour, and record all the cases, including such orders as 1, 2, 3 and 3, 2, 1 as separate ones, you should get to the full 216 ordered cases. This is a mammoth task and best left for a long train journey, or a particularly long and boring lecture (full listing in the Appendix).

Knowledge of the existence of cases in a chance device was of momumental significance in the search for a probability calculus, because, until this could be done and its significance appreciated, there was no way in which a calculus of frequences could be discovered.

Kendall credits the discovery of the full 216 cases of three dice to somewhere in the period between 1220 and 1250 (Kendall, 1956, p. 23), where it appears it was connected with an astrological device for predicting the future of those who consulted it. He reproduces two pages of the 1662 (note that date again!) version of the original manuscript (ibid, pp. 24 – 5).

Problem solving

In the fifteenth century, gaming problems began to circulate, and because they involved gambling it was inevitable, though not too accurate, that the mathematics of probability have been associated with the activities of the casino (much the same mistake is made in respect of early twentieth century developments in statistical theory which have been associated with agriculture, biology – and brewing! – because leading statisticians were researching in these subjects).

Fra Luca dal Borgo, or Paccioli, published a book (1494) which considered the 'division problem' or the 'problem of points'. In this problem, and there were numerous versions circulating, two people agree to play a game until one of them has won six rounds, but for some reason they have to stop the game at the moment when one player has won five rounds and the other player three rounds. The problem posed is: how should the stakes be divided between them?

It would be credulous to consider the problem of points as a subject of enormous practical significance to working gamblers and I think we can discount this as a possibility – it would suggest that games were interrupted so frequently as to require an alternative solution to division by physical violence. The problem of points featured in the literature throughout the next two centuries, suggesting that it was 'merely' a tantalizing puzzle awaiting an abstract solution.

However, whatever its purpose, it signals the beginning of the hunt for a probability of chances, because the implication of the puzzle is that there must, or ought to be, some way of dividing up the stakes fairly, taking account of the chances each had of reaching six clear wins if the game had continued, using as the sole indicator of their chances the evidence of their scores when they stop.

If the division is to be fair, it follows that the information to hand, set against the known possible cases, must be used to *predict* what it is reasonable to assume would have happened if they had not stopped playing (for whatever reason they had to stop – perhaps it started raining, or they were called away on business, or their wives demanded they came home). Paccioli's answer was wrong, and so was that of his critic, Tartaglia, as were the answers of others, until Galileo provided a correct answer sometime before 1642.

The popularity of gaming puzzles continued unabated, and, slowly, ideas about probability spread among the educated. One of the people who thought about gaming puzzles was Gerolamo Cardano (1501–1576). He was, David tells us, a 'physcian, philosopher, engineer, pure and applied mathematician, astrologer, eccentric, liar and gambler, but above all a gambler' (David, 1955, p. 12); in a word, a Renaissance man! He was accused of liberally plagiarizing the discoveries of others and much of the deserved criticism he acquired for this disreputable behaviour was extended, somewhat unjustly, into a criticism of everything else he did.

In addition to the many interesting and controversial aspects of his career(s), Cardano was also the author of the world's first book on probability, *Liber de Ludo Aleae, (The Book on Games of Chance)* (written around 1526, but not published until 1663 – again, note that date), in which he calculated by *theoretical abstraction* that each face of a die has the same chance of being thrown 'if the die is honest'.

His conclusion that there were equal chances for each side of a die is of the utmost importance to the theory of probability. That nobody, as far as I know, worked this out beforehand *and wrote it* down, highlights the significance of Cardano doing so. He did not, we should note, discover this on the basis of his undoubted and well documented extensive experience of dice games, but through the theoretical notion of an ideal and honest die that gives each side an equal chance of turning up.

That games of chance involve the use of a chance device (a pack of shuffled cards, some dice, or a roulette wheel) to produce random events is fairly obvious, but the mere existence of these devices and their organization by structured rules (that is, the game itself) does not of itself create a science of probability, or even, according to the written evidence, an interest in creating one.

Traces exist that the users of these devices knew, at least in a negative sense, that there was such a thing as fair chance, which if left to itself, made the game honest. How do we know this? Simply by the fact that cheating was practised by some gamblers; for if chances were fair it required dishonest intervention on their part if they wished to increase their profits!

For instance, Tibetan priests improved their client's moral stature, and incidently their own welfare, by loading a die (that is, adjusting its shape or weight to favour a particular throw) to land in such a way as to require the priest to recommend a particularly expensive course of religious rites for the luckless individual who had consulted them about some problem or other (David, ibid., p. 10).

If it is known that cheating can pay, it must be assumed they knew that refraining from cheating left the outcome to chance. Cardano, a frequent gambler, presumably knew that cheating was practised in cards and dice games, and he drew the implication from this that a fair game of chance was possible 'if the die was honest'.

Galileo Galilei (1564–1642) wrote on probability on games of chance and solved several problems on it, including one from a correspondent who wanted to know why a 10 could be made by three dice more often than a 9. This was really a problem of combinations of cases, and Galileo found it an easy one to answer. He wrote that there was a 'very simple explanation, namely that some numbers are more easily and more frequently made than others, which depends on their being able to be made up with more variety of numbers' (quoted in Hacking, 1975, p. 52).

The six ways of getting a 9 expand into 25 cases once the order is considered, while the six ways of getting a 10 expand to 27 cases. Where cases have the same chance of appearing, then 10 must be more advantageous for a gambler to throw for than 9 in the ratio of 27: 25.

Exercise 4.4

You could check this if you have completed the exercise suggested earlier of compiling all 216 cases; try it now then look it up in the Appendix).

Galileo's works were published in 1656, in the crucial decade of the birth of probability, and his ideas were more than likely circulated among his pupils and his contemporaries after their gestation.

In Italy, at least, there was some open debate on the various odds that were possible in games of chance but it did not initiate either a research programme or even a substantial literature into the probability calculus.

What caused the birth of probability?

Probability as a subject emerged suddenly. There can be no other word for it. One decade separates the pre-probability era from our own (roughly 1655–1665), and in that decade major steps forward were made in resolving the puzzles of the previous two hundred years, giving a scientific aura to the chance devices of the previous four millenia. Why was this so? Why did not the Greeks, Romans, Italians, Germans, Dutch, British or French discover probability calculus long before, and especially when they were often on the very verge of doing so?

Of course, once the calculus was discovered, it was possible to look back and interpret retrospectively the near attempts at it, which can make the authors concerned look a lot closer to a notion of probability than they were in fact.

Hacking (1975) discusses the various explanations put forward to explain the sudden appearance of probability and his criticisms have a persuasive weight about them. He argues that it could not have been the acceptance of a deterministic philosophy that prevented probability emerging (that is, 'all that happens is intended and is not subject to chance' and so on) because it is precisely in the seventeenth century when probability theory was emerging that determinism flourished. Nor was piety a convincing barrier to a probability calculus because there was a lot less piety about than the literature suggests, and the impious had an enormous incentive to develop a probability calculus: 'someone with a modest knowledge of probability mathematics could have won himself the whole of Gaul in a week' (Hacking, ibid., p. 3).

It was not the technical absence of reasonably perfect dice that prevented the regularities being noticed, because there is evidence that reasonably true dice existed, even in ancient times. The massive empirical evidence of the centuries ought to have counted for something, enough at least to suggest the notion of ideal dice and perfect games. It did not.

Then there is the 'Marxian' explanation: probability emerged when the capitalist system needed it and not before (Maistrov, 1974). Many gifted individuals, however, struggled with the probability calculus for many years before arriving at a solution. Pacioli's probability problem appeared alongside his proposals for double-entry book keeping, and it is true that capitalist businesses and commerce generally require double-entry book-keeping, only he failed to provide a solution to his probability problem of points.

Probability ideas were first applied to the management of annuities for the state and the nascent insurance industry in the Netherlands. Interestingly, the early probabilists got their sums right, but the capitalist managers ignored them and continued to lose money! Whatever the explanation for this neglect of self-interest by people seeking to profit from what they are doing, it must highlight the limitations of a Marxist explanation.

Does this mean that we have no explanation at all for the sudden emergence of the modern concept of probability? As usual the answer is less definite than the question. Probability emerged when it did because it was the natural outcome of the philosophical methods implied in a scientific approach to experience.

It was the application of a philosophical method that did not search for certainty, but only for probability, that distinguishes the pre-probabilistic era from our own. That it is say, the posing of gaming puzzles in itself was incidental to the discovery of probability, for without a notion of probability the puzzles remained unsolved, not for want of arithmetical ability – though that was a feature of their lack of resolution – but for want of the philosophical idea that something less than certain could be defined in quantitative terms of sufficient credibility for them to be the basis for action *or* (*nota bene*) for belief.

It was not Blaire Pascal's (1623–62) solution of his friend's, the Chevalier de Mere's, gaming puzzles (the old teaser of the division of the stakes of an interrupted game and the more complex problem of the real odds for a roll of a double six in 24 throws) that signified the birth of probability, important as these solutions were as demonstrations of the arithmetical competence of the first probabilists. It was the realization that *truth is probable not certain* that distinguishes the pre-probabalistic era from our own.

5

The Definite Maybe

Pascal's wager

Blaire Pascal lived in controversial times. He was intimately involved in a sectarian religious contest within the Catholic church, the details of which are relevant only to students of divinity or the sociology of religion. Religious controversies dominated his short life, though he made major mathematical contributions (in the theory of conic sections), invented a calculating machine that worked, and researched into the theory of gases and pressures.

His letters to Piere de Fermat (1601–65) are often quoted as the beginning of probability – usually with assumptions about the profligate gambling activities of the Chevalier de Mere – but important as these were in spreading the mathematics of probability in respect of the old problems of gaming, they are not enough to explain the philosophical shift from certainty to probability.

For this we must turn to Pascal's *Pensees (Thoughts,* written *circa* 1658, published 1670; 1966). Like Descartes, Pascal had a dream (sociologists or psychologists, or even, perhaps historians, might be able to explain this propensity for mystical experiences among the brightest minds of the age), and in his dream he met and conversed with Jesus Christ. From then on, as a born-again Christian, he was convinced of his life's work – to establish the truth of God. In doing so he contributed to science and mathematics.

Our particular interest is in Pascal's little two page thought, number 418, on the proof of the existence of God (1966, pp. 149

–52), by means of his famous wager. This demonstrates the beginning of probability as a branch of philosophy and mathematics by showing that reasoning about games of chance could be generalized into inferences about other things. The subject of Pascal's wager is of no concern to us, nor are Pascal's deeply felt emotions about what he was doing; progress, as we have noted, uses many strange instruments, and Pascal's wager is certainly among the strangest.

Remember that Descartes convinced himself of the existence of God by the method of axiomatic reasoning – if the premises are true, so are the conclusions; and they remain true even by an act of denial of the premise: to deny the premise ('I think') confirms its truth.

For Descartes, the existence of God was proven with the same authority as the principles of geometry. Reason convinced him, but Pascal rejected a proof of God based on reason: 'Either God is or he is not. But to which view shall we be inclined? Reason cannot make you choose either, reason cannot prove either wrong'.

If reason cannot prove the issue either way for certain we must turn to a method which takes account of uncertainty and yet gives us an unambiguous guide to action (or belief). This is the first significant statement of the philosophy of probability in history – the old era had ended. Pascal argues for belief in the proposition that God exists because of the clear gains from wagering that he does as against wagering that he does not.

His first wager considers the possibility that whether God exists or not; if you bet He does exist and you win, you gain, and if you bet He does not exist and it happens that He does not, then you are no worse off. It is obvious that your bet that He does exist has a positive pay-off if He exists (you gain eternal bliss) and a zero one if He does not (you are dead for eternity) – hence it is rational to wager that He does.

In all cases, it should be noted that the 'stakes' of the wager are the kind of life you lead: if you wager God exists then you live a life of piety, if you wager He does not, then you live in sin.

But, queries Pascal's second voice, suppose in wagering that God exists, which means leading a pious life, I give up more than I stand to gain if I win and obviously much more than I gain if I lose (that is, I have been pious for no eternal benefit at all)? Here Pascal demonstrates his understanding of probability arithmetic.

If the pay-off is the same for an equal chance, it does not matter which way we wager, but if 'there is an equal chance of gain and loss, (and) if you stood to win only two lives for one you could still wager, but supposing you stood to win three' or even more? Clearly, the wager must be taken in favour of the largest outcome. The same applies in this case where there is 'an infinity of infinitely happy life to be won' with 'one chance of winning against a finite number of chances of losing'.

His decision rule is clear: 'wherever there is infinity, and where there are not infinite chances of losing against that of winning, there is no room for hesitation, you must give everything' for you would be 'renouncing reason if you hoard your life rather than risk it for infinite gain, just as likely to occur as a loss amounting to nothing'.

That there are weaknesses in Pascal's argument goes without saying. Certainly, if in a toss of a coin, your opponent pays more on the falls of heads than tails, it is worth betting on heads because the real odds are even for a head or a tail falling uppermost with an honest coin, and over the long run you should win more than you lose. If the life of a dissolute libertine is preferred, it must be remembered that life is finite but God's kingdom of happiness is eternal and therefore the pay-offs are unequal. If the chance is equal that God exists against God does not, the wager is irresistible (assuming you accept the odds and the pay-offs, though if you think an eternity of psalm singing is a dubious pleasure compared to a few years of whatever fantasies move you, it is, of course, not rational to accept the bet).

But, and note this, the format of Pascal's wager is precisely the probability calculus with the notion of expected values of the pay-offs clearly stated, perhaps for the first time in our history. You can reject Pascal's wager and hang on to his method, for that is precisely what marks the birth of probability.

The midwives of probability

Pascal was a member of a catholic sect that supported the

Jansenists, and he and his colleagues became known as the Port Royal logicians, after the enclave in which they worked together. A leading member of the group was Antoine Arnauld (1612–94), who was a theologian and a candid polemicist against the Jesuits. He was one of the authors (Pascal was another) of the group's book, *La Logique, ou l'art de penser* (Logic, or the Art of Thinking) (1662). This book had an impact on European mathematicians and can be said to mark the philosophical birth of probability.

Much of *La Logique* is of no interest to ourselves (earlier caveats applying), but Book IV has four chapters on probability ideas. They signify the first references to probability as a *measurable* concept and the first extention of the ideas of measured probability from games of chance to other events. In effect, the Port Royal logicians went straight from solving the vexed gaming problems of the time to the wider implications of their method to all forms of reasoning and belief. As Hacking (1975) expressed it, before the Port Royal logicians nobody could solve the problems of chance, after them, everybody could.

The authors, for instance, apply probability to the common fear of thunder among the people:

> if it is only the danger of death that fills them with their extraordinary fear, it is easy to show that this is unreasonable. It would be an exaggeration to say that one in two million people is killed by a thunderstorm; there is scarcely any kind of violent death less common. Fear of harm ought to be proportional not merely to the gravity of the harm, but also to the probability of the event, and since there is scarcely any kind of death more rare than death by thunderstorm, there is hardly any which ought to occasion less fear (quoted in Hacking, 1975, p. 77).

If you understand the reasoning in that passage you already understand the basic idea of the probability calculus, namely that you must consider *both* the expectation of an event *and* the probability of the event occurring.

The Port Royal authors understood this point clearly and they hammered it home, applying the method to questions such as that of belief or likelihood, or whether to believe a notary's claims, or whether to believe in miracles, as well as questions about the future. In all these cases it is a question of weighing the evidence, and looking closely at what was most likely.

The *Logic* also included a summary of Pascal's wager on the existence of God, bringing this to the public's attention for the first time (1662), though those readers who rejected the idea of a wager on the emotional (to them) topic of God's existence, also rejected the method itself when, in fact, its implications were much wider.

The Leibniz notation

Gottfried Leibniz (1646–1716), a major contributor to mathematical theory – he discovered the differential calculus in parallel with Isaac Newton – is credited with contributing an important piece of the probability calculus. Leibniz began his career as a student of law and it was in a paper (written at age 19) that he put forward a novel proposition. In studying the law of evidence he was led to conceiving of the relative truth of the evidence as a range running from absolutely untrue, through varying degrees of probably true, to absolutely certain.

Rights to property, for instance, may be absolutely valid at one extreme and absolutely void at the other, or something in between. How do we decide which? Leibniz distinguished between evidence that showed that the claimed right was impossible, evidence that it was certain, and evidence that it was uncertain. In the first case there was no evidence in support of the claimed right and in the second all the evidence supported the claim. In the third and interesting case, the evidence had some points in favour of the claimed right and some against it.

If the claim is absolutely substantiated, Leibniz suggested that this be represented by 1; if the claim is absolutely unsubstantiated it should be represented by 0; and all the cases where the claim is uncertain, being neither totally substantiated nor totally unsubstantiated, it should be accorded some fraction between 0 and 1, depending upon how close it was on balance to one end of the range between absolute certainty and absolute uncertainty. The fraction represented the 'degree of proof' or the 'degree of probability'. In modern usage, this format is used to give the measure of probability, with zero being taken to mean that the outcome is absolutely uncertain, and unity that it is absolutely certain, and numbers (usually in decimals) between zero and unity expressing the degree of probability of the outcome.

Some dicey problems

Antoine Gombauld, Chevalier de Mere, Sieur de Baussay, (a man with three names and two homes, one in Poitou and the other in Paris) flits into the history of statistics on the basis of having posed to Pascal a couple of gaming problems, which Pascal solved to de Mere's satisfaction. The Dutch mathematician Huygens (1620–99) also solved them independently (and, incidently, published a book on probabilty, *Ratiociniis in aleae ludo* 1657).

Interest in de Mere appears to centre on whether he was inspired to pose these questions as a result of (his undoubted) gaming experience or out of abstract curiosity. I think many authors get a vicarious 'thrill' from citing the slightly shady past of an interesting character.

The problem of points, or the division of the stakes of an interrupted game, had a history long before de Mere asked Pascal for an opinion, and, as I have suggested, it has all the hallmarks of a purely technical puzzle posed to test the mathematical ability of someone, rather than a judicial problem arising from a real interrupted game. Poor Paccioli, who first posed the problem (though it may have even earlier Arabic origins), had long gone, and presumably so too had the two players concerned (if they ever existed). That de Mere, 170 years later,was awaiting an answer for a real problem is an unlikely story.

One of de Mere's problems arose from a game that was played with one die. This was (and still is) a real game and it meets my criteria for a good game by being of short duration, with frequent pay-offs, and the possibility of a big score. In it the house bets even money that a player (any player) will throw at least one 6 in four throws of the die. If you can stake as much or little as you like – even a single penny – would you take the bet? If yes, why; if not, why not? (incidently, this is a typical discussion problem given to statistics' students for homework).

To get at this problem, ask yourself what is your bet *against* the house? They are betting that in four throws you will get a 6, and, if you take the bet, you are betting that in four throws you will get any other of 1, 2, 3, 4, and 5, but not a 6.

As long ago as Cardano, we have known that the chances of a particular side of a true die falling uppermost is the same as the chance of any other specified side, that is, one chance in six. Note that we have referred to 'any other *specified* side' and not just to 'any other side'. Why? Because, the chance of any other side, in the sense of any side other than 6, is not equal to that of the chance of the 6 turning up. It is equal to 5/6ths because there are five other sides each with a 1/6th chance of being thrown.

On the first throw you have a 5/6ths chance of not getting a six and a 1/6th chance that you will. If you do, you lose your stake immediately; if you don't you must throw again. The odds are still 5/6ths that you won't get a 6 and 1/6th that you will and they remain so for each single throw of the die. If after four throws you have not thrown a 6 you get your stake back plus the equivalent amount from the house.

At this point you could decide to pocket the cash and leave while you are ahead, or you can do what most winners do and try to increase your winnings. If you lost this first bet, you could also quit while you are only down your initial stake, or you can do what most losers do, try to recoup your losses in a big score (and that, dear reader, could be your road to ruin, though it is exciting while your funds last!)

Why should a gambling house be so generous as to give you an even chance of getting richer, remembering that professional gambling houses do not gamble? The answer is that they are not generous at all; they have better than even odds that you will throw a 6 in four throws. How so, given that at each throw you have a 5/6ths chance against a 1/6th chance that you will not throw a 6?

This is often a most confusing point for students of probability (and people contemplating a gamble at what look like good odds). How can a 'good bet' be to the advantage of the other guy? If we look more closely at the bet we can learn something about the simple rules of probability.

The bet is *not* that on the *next* throw of the die you will not get a 6, but that on *four consecutive* throws you will not get one. It may be that you have a 5/6ths chance of not getting a 6 on the next throw of the die, but what are your chances of not getting one in four throws? It is in fact $5/6 \times 5/6 \times 5/6 \times 5/6 = (5/6)^4 = 625/1296 = 0.482$ (check the arithmetic on your calculator). This is slightly less than 0.5 or 1/2, giving you a less than even chance of winning. Conversely, the chances of you getting 6 must be equal to something slightly more than a half or 0.5. How do we reach that conclusion? We can do this in two ways.

Consider the fact that if out of a certain number of throws, there is a possibility that a prescribed event will occur in some proportion of the throws, what is the probability of its occurrence? Logically, if it occurred on each throw (we got, say, a 6 every time) then the proportion of the event (the getting a 6) equals the number of throws. We would regard the probability of this event as certain, or, in Leibniz's notation, it would have the value of 1. This would require each side of a die to be marked with a 6. If we got three 6s in 6 throws of the die, that is, it occurred in half the throws, we would regard this as a proportion given by 1/2 or 0.5; if it occurred in four throws out of ten, it would occur in 4/10 of the throws (equals 2/5, roughly 0.4); if one in five, 1/5 (or 0.2) of the throws and so on.

It follows that the probability of the event *not* occurring must be the proportion of chances that are left after the probability of it occuring is taken away from unity (which represents certainty). Thus, if the probability of an event occurring is 0.5 it must be the case that the probability of it not occurring is $1-0.5 = 0.5$. You can see this in the case of tossing a coin where the probability of getting a head is exactly equal to the probability of getting tail, or 0.5 each.

If the probability of the event occurring is two in five (2/5 or 0.4), the probability of it not occurring must be three in five (3/5 or 0.6). If this is not the case, what happens in the other three throws? Thus, in arithmetic, $1 - 0.4 = 0.6$. Likewise, just to hammer home the point, if the event occurs one throw in five (or 0.2) it does not occur four throws in five: 1 - 0.2 =0.8. We speak of the probability of an event as being 0.2 and the probability of it not occurring as 0.8.

Going back to our example of the probability of *not* throwing a 6 in four throws, we have found that the probability of this event occurring (that is, no 6 thrown) is 0.482. To calculate the probability of a 6 being thrown we can see that we are dealing with the case where the prescribed event (no 6 thrown) does *not* occur (a 6 *is* thrown). This must be the same as 1 - 0.482 = 0.518, or slightly better than evens. Thus, if the house is offering even money that we will throw a 6 in four throws and we accept the bet, we are working to slightly worse than even odds that we will win.

In the long run we will lose and the house will win. You can see this in the number of times you should win in the original calculation if you were to throw the die 1296 times: you win 625 throws out of 1296, which leaves 671 throws out of 1296 that you should lose (625 + 671 = 1296). Of course, you may be bust long before you make that many throws.

Another problem that de Mere posed to Pascal was much more complicated than this one. He asked why it would not be favourable to a house to offer an evens bet that someone would throw with two dice a double 6 in 24 throws? If, he argued, it is evens or better that a 6 will appear in four throws, with two dice it should follow that a double 6 would appear in 24 throws because the second die adds six times the possibilities of a single die and six times four equals 24.

The answer to de Mere's question goes along the following lines. There are 36 possibilities for each throw of two dice and therefore the chances of getting a double 6 are one in 36 (1/36) and the chances of *not* getting a double 6 are 35 in 36 (35/36). From the method used for the game with one die, if we want to know what the possibilities are of *not* getting a double 6 in 24 throws, we must multiply out 35/36 24 times. To save you the tiresome arithmetic, you can note that the answer is 0.51 approximately. Now this is more than 0.5, and it means that your chances of *not* getting a double 6 are slightly better than even, or, putting it another way, your chances of getting a double 6 are slightly less than even. If therefore you are offered an even bet that you will get a double 6 in 24 throws you should take it, at least in theory (which is after all what the problem is really about) because the odds are slightly in your favour.

Some other simple probability arithmetic

Another simple rule of probability arithmetic is the case where two mutually exclusive events are possible and we want to know what is the possibility of either of them happening. Consider, in the simplest case, the possibility of either a head or a tail being tossed. Clearly, the possibility of either turning up must be certain as there are only two outcomes to a single toss so one of them must occur. In formal terms we are *adding* the individual chances of each event together to produce the possibility of one of them turning up. As the probability of a head turning-up is 0.5 and of a tail is 0.5, added together you get 1, which is certainty.

Traces of probabilistic arithmetic have been noted in ancient Talmudic literature from the second to the fourth century AD (Hansofer, 1967; Rabinovitch, 1969, 1970). The Jews were worried about whether certain acts were forbidden or permitted. The response of the Rabbis, in effect, was to devise a primitive means of establishing the likelihood of an event. In their thinking we can see the simple multiplicative and the additive rules of probability.

Imagine the case of a devout Jew who was walking along a street in a town in which he knew there were nine shops that sold kosher (ritually slaughtered) meat and one that did not. While walking, he found some meat in the street (!) and wanted to know whether he could take it home and safely eat it (on religious not hygienic grounds). The Rabbis answered that he could and they reasoned that as the meat has an equal chance of being from any one of the ten shops, and as nine of them only sold kosher meat, the devout Jew could rest assured that by the weight of the majority he could be sure enough that it was kosher. In other words, they reckoned that a 9 in 10 chance was much stronger than a 1 in 10 chance and they recommended that the Jew acted according to the greater of the probabilities (though they did not use a word like probability).

The addition of probabilities is illustrated in the inheritance case of the woman whose husband died while she was pregnant. Should the executors of his estate assume that a boy or a girl will be born? The ruling said that it was less than even that a boy would be born. How did they decide on this? They argued from the proposition that there were not only two possibilities at the birth of a baby but three: a boy, a girl, or a miscarriage.

The probability that the birth will result in either a girl or a miscarriage requires that the probabilities be added together to give the probability that it would not be a boy. As boys and girls are born, they argued, in roughly equal proportions, once the proportion of miscarriages are added to that of the girls it must follow that the probabilities of it being a boy are less than half, hence, for purposes of deciding on inheritance the executor should exclude the possibility that a boy will be born and, therefore, that the unborn baby was entitled to some provision out of the father's estate.

Another example, this time of the multiplication of probabilities, is the ruling about the fate of an adulteress. Under the law, if her husband declared her taken in adultery he need not support her and could divorce her. Now what if a husband declares his wife an adulteress *before* their marriage is consummated? In the case of adultery before the consumation of the marriage, the Rabbis said that the husband had no case (for divorce that is) even if it is beyond doubt that his wife was not a virgin. Why? Because they said there was a *double doubt.*

The Rabbis argued that it is not certain whether the adultery took place during the year-long betrothal to her future husband or before they met. The Rabbis assumed that the chances were even that it was during the betrothal or before it. They further assumed that there was an even chance that she submitted to her future husband willingly rather than by violence.

As a husband had no claim against loss of virginity before betrothal, he is left with his claim that it took place during it. But the chances are even, or 0.5, that it occurred during it and they are even, or 0.5, that he forced her to submit to him. By considering the two probabilities together they concluded that it was too small for him to have a case. In effect, they multiplied the two probabilities together, $0.5 \times 0.5 = 0.25$ to eliminate the case against the wife.

Towards the definite maybe

Students of mathematics (and not a few lecturers) are liable to be confused slightly whenever they refer to the statistical and mathematical works of *Bernoulli,* for there were eight different Bernoulli's, all from the same family, spread over three generations! Jaques Bernoulli (sometimes called Jacob, sometimes James) (1654–1705) must be distinguished from his brother, Jean (or John) (1667–1748), with whom he quarrelled, his nephews, Nicholas (1695–1726), Daniel (1700–1782), Jean (1710–1790), and his nephew Jean's sons, Jean (1744–1807) and Jacques (1759–1789). They all made contributions to mathematics and statistics.

On this occasion our interest is with Jaques Bernoulli, who was born in the year that Pascal and Fermat were exchanging letters on the arithmetic of probability. He was the author of *Ars Conjectandi* (The Art of Conjecture), published eight years after his death, in 1713.

For Pascal, and the other theorists of probability, the new philosophy was an *a priori* method. They estimated the probability of an event as a ratio between the number of mutually exclusive ways that a single event could occur out of the equally likely mutually exclusive alternative occurrences. Thus, a 6 had a probability of being thrown one in six throws because there was only one 6 out of six possible numbers on a die (that is, the probability ratio of a 6 was 1/6); a jack of spades had a probability of being turned over one in 52 turns because there was only one jack of spades out of 52 cards (that is, the probability ratio of a jack of spades was 1/52 – for any jack, it was 4/52 or 1/13).

The probability of an event was calculated under ideal conditions – no loaded dice, no marked cards. The prediction was *a priori* – before the event – it was not a certainty. Which side was thrown, or which card was drawn in the event, was not and could not be *certain.*

A priori probability is different from *a posteriori* probability (after the event), and Jaques Bernoulli was the first to grasp the significance of this distinction (though Arnauld and Pascal had noted the probability of being killed by lighting by an implicitly *a posteriori* method), and from this distinction, he set out one of the most important theorems in statistics. This section can be regarded as a small salute to Jaques Bernoulli.

An interesting aspect of games of chance is that we can determine at a stroke all the possible cases that can happen (for example, 36 cases with two dice, 216 with three, and so on), and thereby produce a probability calculus, but the facility of easy prediction is also a weakness, as it makes it highly restrictive for application in the wider world. If probability only had to solve gaming puzzles, it would have had little impact, except as a curiosity.

The wider world awaiting is not at all like a game of chance with its finite number of cases; in it there are a multitude of cases with no simple way of enumerating the ratio of what is possible to the full number of those that are available. Bernoulli put it thus:

> Who, I ask, of mortal men will ever define the number, for example, of diseases, regarded as so many cases, which may invade the innumerable parts of the human body at any age and bring death? And how much more easily may one rather than another, plague rather than hydropsy, hydropsy rather than fever, destroy a man, so that from this conjecture may be made on the future state, whether of life or death? Similarly, who will catalogue the innumerable cases of changes to which the air is submitted, so that from this one may conjecture what will be its future state a month from now, not to say a year from now? (*Ars Conjectandi,* quoted in Adams, 1974, pp. 10–12).

Where the number of cases is too numerous to derive predictable probabilities on the basis of definite cases it is essential to derive probabilities by some other means. In this Bernoulli turned to experience, not logic. He ignored the qualms that plagued Hume and others about the reliability of experience as a predictor of the future, and stated boldy and without apology: *experience is a guide to what is probable!*

If the cases are unrestricted, he said we should ask: how the outcomes have behaved in the past:

> If, for example, after an experiment on three hundred men who are like Titius in age and complexion you have found that hundred of them have died before the completion of a decade while the others have lived beyond a decade, you can quite safely conclude that there are twice as many cases in which Titius, too, will have to pay his debt to nature within the next decade than there are cases in which he may pass this limit.

It is as if Bernoulli, on being chastised by Hume for inferring that because the sun rose yesterday it will necessarily do so tomorrow, is replying: 'It (probably) will!' In Bernoulli's (and our) world, it may not be certain that the sun will rise tomorrow, but it is highly probable, and that is good enough to guide our expectations and our actions.

Once Bernoulli released himself from the paralysis of the Humean problem of induction, he opened up an entirely new field of research. It was but a short step, from assigning probabilities to the ratios that were demonstrated by experience, to assert (and then prove mathematically) that the more numerous the experiments the more closely the estimated ratios would come to the true ratio of their probabilities.

This is known in technical language as the *'central limit theorem'*, and Bernoulli was the first to propose it. Now the mathematics of the theorem are outside the scope of this book but we can use an example offered by Bernoulli to illustrate its general properties (note his use of deductive reasoning to give credibility to an empirical method).

Bernoulli surmises that you have an urn (you will come across a lot of urns in statistics), inside of which, unknown to you, there are 5,000 pebbles – 3,000 of them white and 2,000 of them black. The true ratio of white to black is therefore 3:2. How many pebbles, Bernoulli asks, would you draw from the urn (replacing them each time) before you made an estimate of the actual ratio of white to black pebbles? Certainly, you would soon begin to get a rough picture in your mind of the true ratio and your estimate will narrow between some range around 3:2 (say, 303:197 and 299/201) as you repeatedly drew pebbles from the urn (replacing them each time).

Bernoulli's question is the key to his limit theorem. He asks 'can you do this so often that it becomes ten times, one hundred times, one thousand times, etc. more probable that the numbers of whites and blacks chosen are in the same 3:2 ratio actually enjoyed by the pebbles, rather than some other different ratio?'

Of course, if it is not so, he declares, then the experimental determination of probability ratios collapses, but Bernoulli asserts that it is so – the more experiments the more likely that the estimated ratio will approach the true ratio. and in asserting this, Bernoulli is also asserting that the *a posteriori* ratio can be regarded *as if* it was known *a priori,* that is, it has the same degree of certainty as a ratio determined by known limited cases. Though what is probable only has the stature of a maybe, it is, at least, the nearest we can get to a *definite* maybe!

6

On Being Approximately Wrong

Most times we think of errors as mistakes, which is not surprising as it is widely known that those who make mistakes are in error. If, for example, in a futile bid to avoid Micawber's misery of spending more than you earn, you calculate your income and expenditure for the month, and add up the figures incorrectly, you are in error and your mistake can, in principle at least, be corrected, either by checking over your first answer, or by a bank stopping your cheques.

In this chapter, we shall look more closely at errors – both in the sense discussed above and in the sense usually meant by statisticians – and we shall explore the ways in which the inevitable existence of errors has been integrated into scientific methodology. This moves us a little deeper into the more technical aspects of statistical methods, but our approach continues to be gentle and well within the capability of any reader who has got this far. We shall continue to reflect on questions of *why* rather than how as we look closer at the science of approximation.

The use of models

Social science is about people and much of it is written as if the writer, and the reader, can coolly observe what is going on around them without being involved in moral, ethical or political judgements.

Sometimes we increase the illusion (and it is an illusion) of detached observation by modelling some aspect of human behaviour, and isolating from our vision everything considered irrelevant to the particular event. In using models we strip away the non-significant characteristics of the phenomenon and concentrate solely on those aspects of it that we are interested in. In separating the relevant from the irrelevant we are in danger of believing that we can look on the mechanics of the events we are studying dispassionately, like birds circling overhead. This is doubly dangerous, for in stripping away the data that we might consider to be irrelevant we might ignore something very important. We may also forget that we are not capable of isolating ourselves entirely from the mechanics of human behaviour – we are part of the machinery too!

For example, a map is a model of the territory it displays. If it features roads it need not include details of the potholes in them; if rivers are shown nothing need be said about their polluted state, and the contours of hills on a map seldom say anything about the flora on them. Likewise, a model of a new aircraft tested in a wind-tunnel is unlikely to have the seats, luggage racks and drink cabinets placed on board, for these internal details are not considered to be relevant to its external aerodynamics.

Social scientists use a lot of models of the phenomena they study. In management studies, sociology and politics, much use is made of diagrams to describe authority chains in organizations. Economists sometimes use flow charts; the most common being the 'circular flow of income' model taught in the basic courses, which divides the economy into households and businesses but does not distinguish otherwise between or within them. Thus, for purposes of the model the single-parent family on social security is lumped together with the trade union leader who owns two houses and has access to a residential wing of his union's splendid country house 'training centre', and both appear with the middle class family that lives in a 'semi' and have their savings in a building society. On the business side, the sole proprietor is lumped together with the multinational corporation and the nationalized state monopoly.

The model ignores these and other important and interesting aspects of the household and business sectors in order to concentrate attention on the flow of goods and incomes between the sectors. The circular-flow model is a first-blush attempt at identifying the basic relationships of an economy.

A flow model can also be used in sociology to analyse the behaviour of, say, the heroin-using subset of the total population. In this version, the sociologist identifies the eventualities (death, hospitalization, voluntary abstinence or relapse) that are possible for those who experiment with hard drugs. Nothing at all in this model is shown about the personal characteristics of those who take drugs, nor about their motivations.

These personal aspects are not identified in a flow model unless they are regarded as being relevant for purpose of the analysis, such as establishing the proclivities of identifiable groups for particular eventualities – do males relapse more frequently than females, is hospitalization followed by abstinence more often in one racial, religious, income, regional or such like sub-grouping than another and so on?

Models can also be mathematical or statistical in nature, and of necessity these are totally shorn of extraneous data. Mathematics uses single symbols to refer to the population of an entire country, or to a specific group within it. In economics, for instance, the letter symbol L refers to all labour inputs, irrespective of skills, age and the many other differences inherent in actual labourers. The habit of working with highly aggregated categories creates special problems when attempts are made to collect the statistical data that is supposed to correspond to the economist's mathematical symbols, and it is the source of much controversy in and around the discipline.

However, models assist analysis and decision-making – that is why they are used so frequently by scientists. They can describe economically the essence of a complex system and they enable the observer to cope with masses of data that would otherwise be confusing. In addition, good models can suggest simple prescriptive rules for making choices according to the outcomes shown to be available, and for this reason they are widely used in business management. For example, in a model of a company's sales activity, the model might suggest a decision-rule that when sales fall below a given figure, it would be prudent to call in the salesman (whoever he or she is and whatever his or her past sales record) for counselling.

All models that are to be used for description of actual phenomena or for decision-making require data as an input. However, data does not grow on trees, ready for plucking. It has to be searched for, collected, classified and made available to those who want to use it. In the course of this work, those who procure, prepare and present the data can be a weak link in a long chain.

Lies as a source of error

People report on all kinds of things to data collectors – think of how many forms you have filled in so far in your life. It is likely that you will complete many more forms and make many more reports of one kind or another. For instance, in applying for jobs you will complete a *curriculum vitae,* (c.v.), for your prospective employer who uses it to make a judgement about your employability.

Everybody who interviews people for posts and promotions has to learn to recognize that not all people are entirely accurate when it comes to writing their c.v.; not everything in them is strictly the truth, the whole truth and nothing but the truth. Some people are known to fudge 'facts', even if only slightly, in order to present themselves in a better light. They may not intend to deceive, nor may they regard it as wholly reprehensible to do so (some skeletons deserve to be kept in the cupboard away from the prying eyes of anonymous officialdom), but the fact is that personnel specialists act on the assumption that many people fudge their c.v.s and other reports, and this has to be taken into account by those who use such information professionally, and is one of the sensible reasons for interviewing candidates for jobs and asking searching questions of them about their c.v.

If people believe a judgement about them is likely to be made on the basis of what they disclose, they are likely to fudge at the margin those aspects of their life, or life-styles, that they believe will cast them in a less than worthy light (self-esteem being one of Maslow's hierarchy of needs).

For instance, surveys of household expenditure on alcoholic drinks and tobacco are notoriously unreliable as a guide to the actual consumption of these items, whether the data is collected orally from the household or anonymously in a written record that the household keeps of its expenditure over a given time period. We know this because if the aggregate figures for consumption of alcholol disclosed in these surveys are compared with the actual production (net of exports and imports) of alcoholic drink, the gap is so large that we must conclude that most people deliberately understate their consumption levels of alcohol, unless, of course, a lot of alcohol 'evaporates' on its way to the retail outlets!

More seriously, people may deliberately understate their incomes, in order to avoid taxation liabilities. In some countries it pays to be circumspect about one's wealth, and not just for reasons of avoiding income tax; kidnapping gangs are careful to select victims from among the ostentatiously rich rather than the village poor.

Avoiding kidnappers was not the sole motivation for avoiding disclosure of private information, especially when the whole concept of data collection was a novel one and when public authorities were less open to challenge. There was widespread fear of government abuse of the information it proposed to collect and this led to general resistance to the collection of such data when it was first proposed in the nineteenth century.

The suspicions of the cautious were not assuaged by the fact that census and other data of a similar sort was regarded for many years by the authorities as a state secret, neither to be disclosed to the people they governed nor to the prying eyes of other countries, and this fed, to some extent, the paranoia of those who doubted their government's intentions in collecting personal information about them. Sadly, we must acknowledge, that not all such fears are always groundless.

Some people still prefer the state to have less information about the private affairs of themselves than it has today, in case of misuse of that information for political ends. Some think the act of collecting certain information about people has a sinister purpose.

In the 1980 Census in Britain there was some criticism of the question that asked for the country of origin of each person in the household. The grounds of this criticism were that this information in aggregate could fuel extremist calls for discriminatory action against post-war immigrants. The 'immigration' debate is a wholly sensitive subject, made worse, we might note, by ignorance of the facts and the rule of rumour, which, ironically, the census partly intended to correct. Nevertheless, the questions were grossly resented by various organiztions.

In some cases individuals braved judicial reprisals by refusing to complete their census forms, and generated minor flurries in the press and the lower courts as they defied authority in the time-honoured cause of personal freedom. This is usually regarded as typical of the eccentricity that expresses itself in Britain in the great (but lost) battle against such official 'horrors' as compulsory seat belts, decimal coinage, and metrication (surely only in Britain could we find a campaign to 'Defend the Inch'?).

Suspicion of official motives can also be taken to tragic lengths, as in the murder of census officials in Northern Ireland to prevent the chance, however remote, that information collected in this way could be used by the authorities against the IRA.

There are many examples of deliberate misinformation being officially generated for political ends. Government propaganda may overstate the numbers in an enemy's forces in order to bolster the victorious image of their own troops, or to excuse some military reverse. If claimed casualties among an enemy are to be believed, some armies were wiped out several times over (for example, Vietcong battlefield deaths in Vietnam, and mutually claimed casualties in the Iraq–Iran war, 1980–), and some ships have been sunk time and again (for example, HMS *Invincible* by Argentina in the Falklands war,1982).

There are also idealogical biases in the collecting and giving of information. A country that is deeply opposed on religious grounds to certain acts may cause medical staff to list as 'miscarriages' what in another society would be described as 'induced abortions' (similarly, divorce can be described as 'annulment', and loan interest can be banked as 'administrative charges').

People can also be led by their beliefs and prejudices to falsify evidence or to ignore contrary evidence. Years ago I was standing in the main street in Kiev, in the Ukraine, arguing with a Scottish communist about the comparative frequency of public displays of drunkeness in Soviet and Scottish society. She argued strongly that such deplorable scenes as were common in capitalist Edinburgh could not possibly occur in a socialist Kiev. I was vigorously dissenting from such a prejudiced view, when I noticed a man coming towards us, who was not far short of being paralytic with drink. Two women, with difficulty, were trying to carry him between them. I drew her attention to the man and said words to the effect that he was proof positive that Soviet citizens do get publicly drunk. She, however, insisted on being unable to see anybody drunk at all, even though the man and his escorts passed within inches of us!

Ideological bias that leads to the suppression or falsification of evidence is a serious breach of personal ethics, and a great deal of scientific dispute occurs over alleged interference with the evidence by those allegedly embarrassed by it. It is essential, however, to distinguish between the falsification of evidence and our legitimate right to interpret any given evidence differently from our critics.

One way to separate out the reprehensible act of falsification from the legitimate right to differ in interpretation, is for you to publish *all* the evidence you have collected along with your interpretation, and let the readers decide. Unfortunately, it is more common to find disputants selecting (and fulsomely praising) only what they approve of, when, that is, they are not merely blackening the motives and reputations of those they criticize.

Collectors as a source of error

Social science is characterized by the diversity of its sources of data. There are as many sources as there are people alive, and the number of ways that people can be combined together to form groups of data is astronomical. Moreover, as people move about and change so much, it is often difficult (and sometimes impossible) to get a 'picture' of how they are behaving at any particular moment.

The natural scientist can isolate a gas in a laboratory and experiment away for as long as it takes. The gas in the tube today is the same in all essential respects as the gas that was used yesterday, and it is the same as might be used by another scientist in similar experiments thousands of miles away, or even in a hundred years time (I am for the purposes of our discussion ignoring the many important differences between one molecule of a given gas and another which may be of immense significance to a particular scientist). If it was not the same gas, in the sense I am concerned with, the experiment and its results would be invalidated. It is essential for the experimental method in the natural sciences that others can duplicate and check our experiments and derive results that are themselves open to similar checks.

Moreover, the experiment is normally conducted by a trained scientist – otherwise it is unlikely, except in exceptional circumstances, to be recognized as worthy of consideration by the world community of research workers (who can be very snobbish about qualifications). An otherwise unqualified out of work lorry driver, or bored parent, cannot walk in off the street and get a job as a research scientist.

The same cannot be said of those who collect the data for social science. Thousands of untrained data collectors wander the planet compiling the raw material that highly trained social scientists are going to work with. They could indeed be out of work lorry drivers, or bored parents, walking in off the street to be charged with the responsibility of collecting data. What is more, the social scientist has no way of knowing just who collected the data or how reliable they were when they collected it. The only way to be sure is to do it all yourself, which is seldom possible.

In addition, unlike in chemistry, no two groups of people are homogeneous in all respects to anything like the degree of a given type of gas (my caveats above notwithstanding). People change in highly significant and visible ways through time and circumstance – how and why they change fascinates psychologists, sociologists and historians.

This means that the relative certainties of the laboratory are denied to social science, though we need not conclude from this that a data-based social science is impossible and that the only meaningful contribution we can make to economics, sociology or whatever is by way of *a priori* reasoning divorced from empirical data. Agreement that there are no opportunities to *experiment* in the manner of a natural scientist does not preclude the possibility of opportunities for social scientists to *test* their hypotheses about the real world against the available data. Deficiencies in the data are not a valid excuse for ignoring data altogether.

Experienced research workers usually get a feel for the reliability of the particular data they use after some exposure to its idiosyncracies. This teaches them to use all data carefully, but it does not necessarily prevent them using whatever is available in spite of its known (or suspected) deficiencies, for they must use inadequate data if there is no other data available or likely to become available (though they do draw their readers' attention to its defects).

For example, those who need to work with statistics produced by the United Nations and its agencies learn to distinguish between the reliable and the less reliable entries in the UN tables. The deficiencies in UN data are not the fault of the UN; it should be noted that UN statisticians have done as much as anybody to standardize and improve data collection and presentation.

It should be remembered that UN statistics are supplied by the member governments, and generally are accepted at face value by the UN – the diplomatic consequence of accusing a member state of 'fiddling its books' is believed to exceed the likely scientific benefits of doing so! But the point must be made that poorer member states of the UN obviously do not have the same quality of data collecting facilities and statistical resources as the richer and better endowed members. Some are deplorably deficient in such matters and others deliberately neglect them. Hence, UN data across 140 members is not strictly comparable and no research worker can be expected to know the extent of the wide variations in standards of collection, though it is not uncommon to see papers quoting UN data to two or more decimal places!

At a local level, the deficiencies in the collecting staff are serious. Many enumerators are untrained volunteers earning some extra cash for a few hours work. In statistics, the use of the untrained enumerator has a long tradition, going back at least the 'searchers' who collected information for the Bills of Mortality that John Graunt used for the first statistics.

Training in itself does not remove the problem, though it certainly reduces it. The use of survey methods has become fairly common over the past thirty years, and great improvements have been made in designing questionnaires and in conducting the actual surveys. Unfortunately, individuals can read the same question differently, just as two actors will differ in their interpretation of the same lines, and the variations in tone, clarity and mood can influence the answers they get from the interviewee.

If the interviewer implies that a certain answer to a question will be regarded as evidence of moral depravity, this might suggest to the interviewee that discretion is preferred to honesty. This could explain why in opinion polls the extent of the support for extremist parties or extremist measures is often underestimated, compared to the evidence of secret ballots, because people may be deterred from publicly admitting their support for extreme positions if the questioner implies such behaviour is disreputable.

Leaving interviewees to read the questions themselves may not overcome this problem. The questions may be ambiguous or may not have categories that include the particular interviewee's behaviour. Written questions can imply prejudices, even when not intended and they can lead to misleading answers.

Mistakes can be made, and not corrected, in completing a questionnaire. The enumerator can mishear an answer or tick the wrong box through carelessness. Attempts to correct an error can mess up the page to the confusion of the clerk who checks through it, often at speed, and render useless the response. The number of cases of obviously nonsensical responses – 'widowed 12-year-olds', 'male grannies' and 'unemployed deceased window cleaners' are among the legions of census howlers reported to punch operators during their training – and suggests that enormous significance ought not to be credited to everything that purports to be a survey.

Data as a source of error

Returning to the example of the scientist experimenting with a gas, we can assert confidently that references to the gas do not cause ambiguities among other scientists – they will be expected to know precisely what it is that the experimenter is referring to in any papers published on that particular gas.

It is not quite the same in the social sciences, for they abound in vague, not to say, positively confusing, concepts and categories, and yet, people act and react to propositions regarding these concepts and categories as if they had the precision common in the natural sciences.

Take, for instance, the everyday subject of price. Almost everybody copes with the concept (and, if in funds, the actuality) of prices. But is the term as unambiguous as it appears? Suppose you were studying the movement of car prices since, say, the 1973 oil price hike. What would you enter in your survey as the price of a car? Is it the ex-works price charged by the manufacturer to the dealer? If so, which of the ex-works prices would you choose? The main-dealer price or the one-off price for a small garage? The discount on the ex-works price could vary according to the annual sales of cars at that outlet. Does that price include delivery by the manufacturer to the buyer's premises or is it the factory gate price plus delivery?

It is no less confusing at the consumer end of the price chain. Showroom car prices entice the customer in for a closer look. But is that *the* price of the car? If not, what is its real price? Anybody who has dealt with a car salesman will know that the displayed price need not be the actual price on the bottom line of the contract. If the salesman needs to sell his cars more than the customer wants to buy them, his prices will fall to 'never to be repeated offers', and if the urges to buy and sell are reversed, the salesman's price will rise as he adds on the 'extras' ('delivery', number plates, and so on).

Similar variations in meaning can occur with other everyday concepts in social science. Unemployment ought to be a fairly straightforward category, but is it? In measuring unemployment are we measuring all those who are available for work but without it? This could include those people on the unemployment register and also many others who have not registered for one reason or another, such as housebound parents who have not bothered to register, or people only looking for part-time work, or people unable to register because of residential disqualifications (illegal residents).

The unemployment register itself may not survive close scrutiny as a reliable indicator of the unemployed (that is, those out of work and looking for it). It can include people who register to receive the benefits but who have no intention of accepting work if it is offered. Others may be unemployable, in the normal sense of the term, for reasons of ill-health, poor physical condition or personality disorder. Some people might also be unemployed only by virtue of their unwillingness to work for the wage levels on offer, and meanwhile they are 'waiting' for wage levels to rise.

If we define unemployment then as the numbers not working 'full-time' (how do we define full-time employment?) out of the population between the age of 16 and 65, we would get a different (higher) figure than we would by taking only the registered unemployed.

Each country measures its unemployment by different criteria and uses different methods to do so. In the United States they use a survey method – much like an opinion poll, except instead of asking who or what they support, they ask if they are working or not working – and in the UK the unemployment figures are collected from those on the weekly registers of the unemployed around the country. There are many other complex problems in measuring unemployment rates in different countries, which excite specialists who study them and report to government agencies, but we have said enough to show that the simple concept of unemployment is a less 'obvious' category than it seems at first glance.

We could elaborate on the complexities of defining common data and the consequences of these complexities for the social sciences but enough has been said to give you the flavour of some of the technical controversies that beset social scientists whose work is based on such data. We can sum this up by asserting that errors arising from the confusion of definitions are as serious as those arising from the other causes we have mentioned, and that in consequence much effort has been, and is, expended by various groups of specialists to establish more rigorous definitions of categories and concepts and to enforce minimum quality standards in the collection and presentation of the data.

This last creates another, sometimes major, problem in statistical work. Attempts to 'improve' the data base – either by clearer and stricter definitions or by more accurate standards of measurement – have an obvious positive effect, but they also can have the negative effect of discrediting previously collected statistics, or of, at least, making them unusable in conjunction with newer data.

The result is annoying breaks in the time series of many collections of official data, because the previous data is now incompatible with what is issued subsequently. If this happens too often (as, alas, it does in some official series; note in this context that the method of compiling the UK unemployment figures was substantially changed in 1983) we end up with runs that are too short for useful work.

Therefore, when we read of controversy in social science we ought first to check that the parties are using comparable data; if they are not, it suggests that they may be like two scientists arguing about which of them has made the correct diagnosis of *different* chemicals, while believing them to be the same.

The notion that such an error would be rare, if not impossible, in the natural sciences (in the sense that its occurrence would itself be the cause of severe censure on the participants for being so utterly 'unscientific' as to use the kind of sloppy research methods that they cannot distinguish between essentially different chemicals) ought not to invalidate the fact that similar errors of definition and measurement are all too common in the social sciences.

7

On Being Approximately Right

The futility of perfection

I do not mean to give the impression that the natural sciences are devoid of error, either in the form of mistakes, or in the statistical sense that there is inevitably some degree of error when humans handle data. It was the recognition that errors were present in even the most carefully conducted experiments, or carefully managed observations, in the natural sciences that gradually led scientists to take account of error in their work.

Certainly, the Greek astronomers were aware of discrepancies in their observations. Their data base for the study of the movements of the stars, sun, moon and planets was long enough to provide evidence of a lack of perfect regularity in the length of an Earth year. The Greeks knew that the year was not exactly 365 days long, and Ptolemy calculated it to be $365 + 1/4 - 1/300$. He did this by comparing observations by the Egyptians, the Greek astronomer, Hipparchus, and himself.

In astronomy, a large number of people observe the same objects – they do not mistakenly look at different heavenly bodies. The history of astronomy is largely about achieving greater and greater accuracy in observing the same objects by means of improvements in the telescope and other related apparatus. It is also about improving mathematical capabilities and techniques. In combination, apparatus and mathematics made astronomy into a science, and from this came the discovery of the simple arithmetic mean and also the more mathematically complex normal curve.

As the numbers of observations increased, so did the number of small variations in the readings noted by the observer or observers. Some way had to be found for reconciling discrepant observations, other than by attributing them merely to the observer's degree of care or to the quality of his instruments. Both sources of error were slowly reduced as individuals practised being more careful and instruments were gradually improved.

However, it is not possible in practice to eliminate errors in measurement completely. You can test this yourself by taking a ruler and measuring the page size of this book. Repeat the measurement several times, each time removing the ruler, and note down your answers. There will be small variations each time you do this, assuming you are a normal person.

The existence of small errors in astronomical observations were more than an irritation to perfectionists who sought the same measurement for every object every time they glanced skywards. Small discrepancies could have important implications for astronomy (and for the practical applications it had for navigation).

The astronomer Tycho Brae calculated 'averages' of his observations to get a 'truer' result and this method was widely known by the early seventeenth century. In principle it seems logical to do this. If two observations, taken in broadly similar circumstances, are added together and divided by two, we get the average of the two results and this is more representative of the true measurement, if the errors cancel out.

Brae, for example, in a series of observations of the ascension of a star, Alpha Arietis, over a period of six years, found the average of pairs of observations of the star in relation to the planet Venus. Each pair of observations was added together and divided by two to give an arithmetical average, and this average was noted. Thus, for instance, he took the measurements for February 27 1582 and September 21 1585, when Venus and the Sun had roughly the same altitude, declination and distance from the Earth, and found them to be 26° 4′ 16″ and 25° 56′ 23″, respectively. He added them together to get 52° 0′ 39″ and divided this by two (there being only two in the group) and got 26° 0′ 20″.

In all he did this for 12 pairs of observations, plus three single observations, and in this way averaged out the discrepancies. One result of this method was to reduce the range in the observations from 16′ 30″ (the lowest reading being 25° 52′ 22″ for December 27 1586, and the highest 26° 8′ 52″, for November 29 1588) to plus or minus 6″.

Pairing produced average readings much closer together than the spread of the single readings themselves. His last step seems to have involved a calculation of the arithmetic mean of all his averaged results to produce a value of the ascension of this star to within 15 seconds of its modern calculated value for 1585 (Plackett, op. cit.,p. 123).

In this case, the use of the arithmetic mean helped to cope with variations in observations but the application of the arithmetic mean need not be confined to problems involving discrepancies in observations. It can also be used to find the 'representative' value of a common set of data – the 'average' starting salary of graduate teachers, or the 'average' length of service of a group of employees, and so on. In this case the differences in the data are not due to errors of observation; they are real and definite differences – some people are paid more than others, some people have served longer than others and so on. It is a *representative* (or descriptive) value we are after, not greater accuracy.

Almost certainly, every reader knows how to calculate the average of a set of numbers, say, the average age of the people in their class, or the average income of the people in their household. You simply add the ages, or incomes, of the people concerned and divide the total you get by the number of persons. The result will be the average age, or income, of the group.

Exercise 7.1

You could try an arithmetic mean, or average, by counting the number of words in each line on this page and finding the average number of words per line; you could also check your answer, roughly, by multiplying the average words per line that you find by the number of lines on the page and comparing this with the total words on the page found by summing the words per line (see Appendix).

You should note again that statisticians have a special name for what I have referred to above as a 'group'. They refer to the people, units, or items that make up a set or group of data as a *population*. By this term they do not confine themselves to human beings; in statistics the population can be anything – cricket bats, or nuts and bolts for instance. You will slowly get used to this peculiar use of a word that up to now you may have associated only with numbers of people (jargon has a habit of growing on users!).

Some simple experiments in finding the arithmetic means of various populations would soon show you that it is not necessary for any particular item in a population to have exactly the average value of any particular characteristic – the average age of a family might be, say, 25.5 years, while everybody in the family might be either under 20 or over 40. The average is a computational value more or less representative of the population to which it refers. How an average can be made more representative is discussed in the next section.

Frequency

The calculation of the *arithmetic mean,* produces a number, more commonly known as an *average.* You will also see this referred to as a measurement of central tendency. All the values in the particular population are included in the calculation of the arithmetic mean and therefore they all have some effect on its value. This can lead to unrepresentative averages, where the majority of members of the population are not represented by the calculated average – some have values way above, and others way below, the average or mean value.

Thus, if we are looking for the average wage of a population of employees, we are bound to get a different average wage if we include in the calculation the wage of every employee from the junior clerk through to the managing director, instead of confining the population to, say, the wages of the blue-collar employees only. The range of wages influences the value of the average wage.

If the managing director, and the management employees close to him in income, receive considerably more than the blue-collar employees in the workshops, and we include them in our calculations, the resultant average wage we calculate for the employees of that corporation might be very much higher than the wage that any blue-collar worker actually receives.

If, however, we limit the population to the blue-collar workers only, and find their average wage, it is more likely to be close to and representative of the actual wages of that particular population of employees.

How data is selected can have real importance when a corporation and a union are trying to negotiate a 'fair' wage contract. The corporation will claim that average wages are relatively high in their plants compared to others, while the union will claim that they are relatively low.

Even if the ploy of including the entire workforce in the calculations is too obvious to be successful, there are other ploys open to both sides to try on. The corporation attempts to boost its negotiating position by counting all earnings, including overtime, bonus payments, shift premiums, pension contributions, fringe benefits and such like as part of average *earnings* of employees; the union tries to exclude everything from the calculation except the basic hourly rate (which few members actually earn) and uses this to prove that the corporation are a 'penny-pinching bunch of slave drivers'. Other ploys are also obvious and are frequently used.

The union concentrates its case for a wage increase for all members on the plight of the least skilled and lowest paid group. In appeals to the public for support, it produces pay-slips from unmarried employees who have the largest tax liabilities and therefore the lowest net wage or take-home pay. It also prefers to talk about hourly rates only in all discussions. The corporation reverses this ploy in an effort to boost take-home pay, by talking about the incomes of their married skilled employees, and it always talks about gross, and preferably annual, earnings. The intent is clear: include a wide range of high value data if you want to 'raise' the average; exclude high value data if you want to 'lower' the average.

If you look closely at the arithmetic mean of a given population (in our case, the hourly rates for a group of employees, all with different skills, different extra payments and bonuses and so on) you will notice that there is a difference between the earnings of each individual in the population and the arithmetic mean earnings of all (except, of course, for those that have the same earnings as the mean). These differences are very important and they have an interesting property. (If, however, there are no differences in the earnings of a group of individuals, then the average earnings would, of course, be the same for every individual, but if this was the case there would be no point calculating the average!)

If you add up all the differences in earnings between each individual above the value of the mean earnings, counting those values identical to the mean as zero, you will find that this total corresponds exactly to the sum of the differences in earnings of each employee below the mean. In other words, the sum of the positive differences above the mean is equal exactly to the sum of the negative differences below the mean, or, in technical words, the algebraic sum of the differences is equal to zero.

It is possible to use the property that the deviations of each individual from the mean sum algebraically to zero to find the arithmetic mean by another method to the one most commonly used. In this method, you 'guess' the mean value of a set of data and add the algebraic sum of the differences or deviations of the data to your 'guessed' or assumed mean.

If your guessed 'assumed mean' is higher than the true mean, the algebraic sum of the differences would be negative (there being more values of individual differences below your assumed mean than above it) and when this negative sum is added to the assumed mean it reduces it exactly to the true mean; if, on the other hand, your guessed assumed mean is lower than the true mean (there being more values of individual differences above your assumed mean than below it), the algebraic sum of the differences is positive and this raises the assumed mean to the true mean. Calculating averages by means of the assumed mean method can sometimes seem to be dry as dust, especially when done the long way with pencil and paper instead of a calculator, but you should note, as you plough through the calculation, the concept of individual values in a population deviating from the mean, or average, value of that population.

This will help you to understand, and therefore to tackle, the topic of *frequency distributions* using *grouped data.* The arithmetic of the calculations is less important for us in this book than the statistical concepts involved. This method enables you to calculate a more accurate (in the representative sense) arithmetic mean of a population, and it might help settle some otherwise difficult arguments in social science and management studies (made more difficult where there are conflicting interests in the outcome).

Consider again the dispute between a corporation and a union over the money value of a 'fair' wage increase. If the current average wage includes not only the wages of the union members, but also the higher wages of the management, it is going to look a lot higher than it would do if the wages of union members only were included. We can illustrate this with a simple example. If the corporation paid its employees in the blue-collar grades £1,000 a month, the management grades £2,000 a month, and the managing director £4,000 a month, you could work out the average of the wage levels paid by the corporation by adding the three wage levels together and dividing that sum by three. This gives an arithmetic mean wage of £2,333 a month. This would probably surprise the blue-collar employees as it exceeds by £1,333 their own monthly incomes!

The union would no doubt protest at this way of calculating the average wage and claim that it produced a phoney average totally unrepresentative of the actual wages earned by its members. The union might argue that weight ought to be given in the calculation for the large number of employees (2,000, say) who are on lower wages compared to the number of managers (151, say) on higher wages. In effect, they want to calculate the average so that the numbers earning each wage level influence the calculation of average earnings. We can do this by using the frequency distribution method.

We can illustrate the principle involved in this method. Briefly, we multiply the number earning a given wage by the amount of the wage, and sum the total to get the monthly wage bill. Thus, there are 2,000 blue-collar employees earning £1,000 and this costs £2 million; 150 managers earning £2,000 cost another £300,000 and finally the single managing director at £4,000 a month costs £4,000. This gives a total monthly wage bill of £2,304,000. We want now to find out how the total is divided up on average between the various grades of employees, while allowing the blue-collar grades to weight that average by virtue of their larger numbers.

The weighted average is found by dividing the total monthly wage bill by the number of employees (2,151), and this gives an average wage per employee of £1,071. Now this average wage is still greater than the £1,000 a month wages of the 2,000 blue-collar men, but by only £71 a month instead of £1,333. The effect of calculating it this way (known technically as the *grouped frequency distribution method*) has been to increase the influence of the large numbers at the lower end of the wage distribution, though not by enough to satisfy the union entirely. The union has merely discredited the management's version of the average wage in the corporation.

The £71 difference means that the union can still cry 'foul'; but using a frequency distribution the average is at least more closely representative of what the majority are paid, than the case where (without taking account of the numbers receiving each level of wages) a straightforward average of three wage levels is used.

We ought to note here that there are two other measures of the average: the median and the mode. I will leave you with a statement about each and press on with the next section.

The median is found by ranging all the values in ascending order and counting until you are exactly half way through the number in your population. The wage level at that point is the median wage. In our case, there were 2,151 employees and counting along until you found the 1,075th employee would give a median wage of £1,000 (wild cheers from the union side, glum faces from the management).

Hence, in the median measure, exactly one half of the cases are counted and the value of that case is taken (though, in practice we would have to interpolate between two values at the mid-point if there is an even number of units in the population).

The median is representative and eliminates the influence of extreme values – if the managing director earned ten times as much as he does in our example, it would not influence the value of the median.

The mode is slightly different, though in our example, it may not seem to be. In this measure, the value of the maximum number of cases in the group which have the same value is taken as the representative value. Thus, we have 2,000 employees on £1,000 a month and 150 on £2,000 and only 1 on £4,000. The majority earn £1,000 and this would be regarded as the mode for that population. In this population, the mode and median coincide (to the great relief of the union). You might wonder why the mode is regarded as being useful and it must be said at once that there are cases where it is less useful than the other measures.

In the main, its most useful role is in qualitative rather than quantitative descriptions of data. Consider the case when we are looking at the popularity of different subjects taken by students. Which is the most frequently taken class? Clearly, the one in which the most students enrol. This is the mode. It would not mean much to use an arithmetic mean as a measure here – how do we average across different classes?

You should note in conclusion that the measure of an average is a number that is meant to be representative of the population to which it refers. Some averages are more representative than others and we must take care when handling averages not to confuse precision of the number with representative accuracy. How representative a particular average value is of a population will depend on the range of values across which it is calculated – the wider the range the less representative an average measure is of its population. This statement is so important in statistics that it is worth repeating: the *wider* the dispersion of values in a population, the *less* representative the average value of the characteristic being measured is likely to be of that population.

If we calculate the average age of the students in a university class, it is likely that the dispersion of actual ages about the average age would be less than the dispersion of actual ages in the case of, say, a three-generation family, given that the majority of people at university would be of similar age, which is not (legally) possible for a normal family (parents by law, and to some extent by biology, having to be at least 16 years older than their children!).

The subject of the dispersion of the data about the value of the arithmetic mean is one of the most important recurring topics in statistics and is the source of much popular misconception. It is worth considering here.

On deviance

At the turn of the century, sail had almost given way to steam, though some highly experienced captains continued to ply their trade with ships past their prime, often in the most astonishing manner. One such captain was William Andrew Nelson (1839–1929), from Maryport in England. In his sea career he sailed over 100,000 miles, often round Cape Horn. During this time, apart from not losing a ship – itself a feat in those days in those ships – he managed to achieve best and worst times for all his voyages that did not differ in duration by more than 5 per cent. This is an astonishing achievement and one that must have made him attractive to numerous owners who were looking for someone totally reliable, even though other captains sometimes had rare flashes of luck and beat his best time for a particular voyage.

Suppose that Captain Nelson's average time for a voyage to California via the Horn was 118 days, and that another captain also averaged 118 days. Which of them would an anxious owner want to hire, given that they both had the same average sailing times?

We know from Captain Nelson's logs that his worst time for the voyage to California would not be longer than 5 per cent of his best time, and if his best time was 116 days, his worst time would be about 120 days. Hence, his average time of 118 days could have a range either side of it of only 2 days.

What of the other captain's average of 118 days? This could be composed of voyage times running (on two occasions) as low as 112 days and (on one occasion) as high as 125 days, with other times around 121 days. His range is greater than Captain Nelson's – up to six days each side of his average of 118 days.

Most owners would be likely to prefer Captain Nelson for a voyage, as the range is smaller around his average time. Similarly, if we are told that the wages on offer were above average we would like to know by how much they were above average and what was the range of wages either side of the average. Suppose the wages on offer were £500 a week, while the average wage for our kind of talents was £490 a week. Would we be delighted? Not if we knew that the lowest wage for our trade was about £280 a week while the highest was £900, though we might have a different attitude if the range around the same average was between £485 and £500.

The range between the highest and the lowest can vary enormously and we ought to take that into consideration – indeed, the variation about the mean can be of importance in making decisions. I have introduced the idea of the range between the highest and the lowest values of a distribution and suggested that this give us a clue as to the relative attraction of a particular place in the range relative to the average value of the whole population, but it would be much more convenient if we could find some measure of our place in the range that does not require us to examine the entire distribution everytime we consider the data.

One such method is to find the average of the deviations of each item in the population from the arithmetic mean, taking care to neglect the negative signs on those deviations below the mean (otherwise we would end up with a zero when we add them to positive signs of those values above the mean). The arithmetic is the same as for any other arithmetic mean – add up all the deviations and divide by the number in the population. Intuitively, this measure of the average deviation has a simple appeal, it uses the concept of the average for instance, and we like best to work with the familiar.

The resultant statistic – the average deviation – is descriptive of the range of the data. If the average deviation is relatively high, it suggests that the range is widely spread around the average for the sample; conversely, if it is relatively small, it suggests that the range is relatively narrowly spread around the mean.

It is more common, however, to take the *squares* of the

deviations about the mean (that is, multiply each deviation by itself), as this automatically eliminates the negative signs (a negative multiplied by a negative is always a positive) and then to divide this sum by the number in the population. This measure is known as the variance and is designated by the Greek letter sigma, σ^2 (read as 'sigma squared'). (Incidently, statisticians also use the Greek letter μ, read as 'mu'. for the arithmetic mean of a population.)

Squaring the deviations of large numbers and summing them leads to quite large totals, which are not always easy to interpret at a glance. Certainly, you can compare two variances at a glance and conclude that the smaller one represents a smaller deviation of its population about its mean than the larger one, but you will not be able to do much more than that.

For many uses, a more representative measure of the deviations about the mean is the *standard deviation*. In the standard deviation measure we not only square the deviations about the mean, sum them and divide by the population, as we do to find the variance, but we also find the square root of the answer. The result is known as sigma (σ). This gives us a smaller number, no matter how large the value of the variance, and we can compare it directly with the size of the average for that population.

When you see the standard deviation, compare it with the number given for the average. This gives you a simple way of judging how representative any average is of its population: in general, the smaller the standard deviation compared to the size of the average, the more closely the items in the population are clustered around the average, and, therefore, the more representative the average is of that population; conversely, the larger the standard deviation compared to the average, the wider is the dispersal of the deviations of each member of the population around the average, and, of course, the less representative it is of its population.

At a glance then, the standard deviation gives us a measure of the relative deviations of the items in a population about their arithmetic mean. You can see the significance of the standard deviation measure by considering some examination scores. If, say, the average pass mark was 58 and your mark was 62, you (or your examiners) would be able to form a judgement of how well you did – apart from having passed rather than failed the examination – by calculating the standard deviation about the mean of the pass marks of all the students who sat that examination.

If the standard deviation is calculated to be 8.1, then your pass of 62 is less commendable than it would be if the standard deviation was 1.8; in the former case the large standard deviation suggests that the bulk of the class scored between 50.9 and 66.1, which makes your mark of 62 OK, but not outstanding; in the latter case, the smaller standard deviation suggests that the bulk of the examinees scored between 56.2 and 59.8, showing your pass mark of 62 to be comparatively better than their's, making your result commendable.

Measures such as the standard deviation tell us something about the *dispersion* of the bulk of the items in a population about the arithmetic mean. That we can measure how a population clusters around a mean is interesting in itself, for reasons discussed above, and would be worthwhile if that was all we could do with the standard deviation. However, there is much more for us to get from measures of the dispersion of a population about its mean value and we shall continue to explore this topic here.

Clustering about the average

In a famous dispute in astronomy in 1572, the issue in contention was whether a star that had flared up was nearer to or further from the Earth than the Moon. This remarkably naive debate has to be placed in the context of the sixteenth-century belief in Aristotle's astronomical system, which conceived the outer celestial system to be eternal and unchanging. The flaring of the star in the Cassiopeia constellation was evidence that change was possible in the outer system, unless, according to the beliefs of classical astronomy, the star was closer to the Earth than the Moon. We know today that such a position is impossible for a star, but the early astronomers had no good reason to think so themselves, unless they could prove it scientifically. Much controversy raged among them about the implications of the star's behaviour, which did not end when its brightness diminished and then disappeared in 1574.

Galileo joined the debate and in doing so he introduced a theory of errors to the experimental method. Galileo criticized the belief that the star in question was between the Earth and the Moon. He noted that the various observations made at the time showed the star to be different distances from the Earth.

The differences in the measurements, said Galileo, were caused by errors in the observations, and these errors were 'absolutely unavoidable'. An observation 'with the same instrument, in the same place, by the same observer who has repeated the observation a thousand times' will result in a variation in the reading of 'one, or sometimes many minutes'.

How then could scientists cope with these observational errors? By ignoring the largest errors and concentrating on the minor ones, said Galileo, for the former are rare (and obviously wrong) while the latter were frequent and likely to be contradictory in algebraic sign. Errors, asserted Galileo, were 'equally prone to err in one direction or the other'. That is, if there are errors in excess of the true value of a measurement, there are likely to be similar errors in the other direction and these errors would tend to cancel each other out to produce the true (or truest) measurement. This idea that errors in the observed measurements are *symmetrical* around the true measurement is similar to the idea, discussed in the previous section, that deviations are distributed around the mean.

If the observer stuck to those measurements that clustered closely round some value – while ignoring those others that were substantially different and obviously wrong – he would get a very good approximation of the true value, Galileo asserted. In fact, the greatest number of observations would cluster around the true value and only a few 'maverick' ones would be a great distance away.

The rationale for calculating the average value of all the observations is clear: as the measured observations would cluster around the true measure, and the greatest number of them would be closer, rather than further, from this true measure, an average value of all the measurements would be heavily weighted by the bulk of the measurements and thus the arithmetic mean would be very close to the true measurement.

How close the arithmetic mean was to the true value of a measurement would depend in part on the accuracy of the observer's measurements and in part on the number of them; the former being a function of the standards of scientific rigour attained by the observer and the equipment, and the latter a function of the observer's patience in collecting data.

Once stated, this simple notion of averaging the observed measurements to approach the true measurement of something, obviated becoming paralysed by a hopeless quest for perfection. Being absolutely right was not considered feasible for scientific practice – perfection, like utopia, is not attainable – but it was possible, instead, for precise ideas to be generated about the probability that the average of carefully made approximate measurements were as close to the true value as it was possible to get.

We shall now turn to the most important single concept in basic statistics, namely that of the *normal curve.* Nothing is quite as crucial to statistical theory as this particular mathematical construction, and nothing is so widespread in its application.

As usual, the best way to approach the normal curve is from a historical perspective, to show, first, what the curve attempts to explain, and then to indicate an early application in the social sciences. I shall not take a strictly chronological route, for I intend to look at an application of the normal curve to social science in the nineteenth century, rather than take you through its mathematical generation from De Moivre in 1733.

The average man (or woman)

Adolphe Quetelet (1796–1874) was a Belgian. His first ambition was art, particularly painting, and he also published a little poetry. To makes ends meet – for those with pretentions to art in his day truly suffered in the proverbial garret – the 18-year-old Quetelet took part-time work as a teacher of mathematics. Seldom does a chore lead to a creative life's work, but he was persuaded to undertake a PhD in mathematics. At age 23 he completed his thesis on an abstruse theorem in analytical geometry, and went on to become a teacher of mathematics in Brussels, at the behest of the Ministry of Education. His love of art and art forms did not whither and he continued to write about art with the passion that we normally reserve for our first love.

From here on, Quetelet's life lines began to converge. He undertook the responsibility for establishing an observatory in Brussels (why is not known) and as a result he was sent to Paris to study astronomical methods in 1823. At this time the mathematics of astronomy were well into probability concepts and the theory of errors in observations. Quetelet took to the methods of astronomy with a gifted sense of application and he combined them with his first love, art. For one thing, he noted that the human body was not unlike the observers' universe in its minor diversities of form and simultaneous similarities. The notion that the human form was diverse, yet similar, was probably obvious to art students who looked at and drew, painted or sculpted enough human forms to be sensitive to diverse similarities – but until Quetelet, the notion had no scientific interest.

Look about you on any crowded beach at the height of a hot summer and you are bound to agree that the human body comes in all sorts of shapes and sizes. There are four extreme shapes and sizes: emaciated or gross; midget or giant, and a notional 'ideal' form for the human body (though ideas about what is ideal vary across the ages, as a history of art shows). The majority of people range between being too thin or too fat, too tall or too small, too 'top heavy' or too 'bottom heavy', and numerous combinations of each characteristic. If they are at an extreme in one dimension they may be closer to the norm in others. In short, there is an average shape and size in the human form.

The artist in Quetelet grasped this at once and linked it statistically to a mathematical explanation of it for the first time. What had always been obvious, though unremarkable (in the 'so what' sense) to the artistic observer of the human form, was given a scientific meaning: just as observations tended to cluster around the true mean measurement, so human shapes and sizes clustered around the true mean shape and the true mean size, making the majority of humans less than ideal in form. Nature allowed variations from its original design, some of which, in a relatively few cases, were quite large, while for the the majority of cases, the variations were minor.

Quetelet thrust himself into the work of collecting data on the various characteristics of the human form and the behaviour (largely criminal) of segments of the population. Between 1826 and 1840 he published numerous studies of the data on the average height, weight and criminal tendencies of the Belgians and showed how they clustered around mean values – the majority of people were broadly similar in height and were also broadly law-abiding. The exceptions were a minority of cases that got scarcer as the extent of their deviation differed from whatever was the norm.

He was on the verge of his greatest discovery, only requiring to go beyond the calculation of averages for his data – the average height, weight, and propensity to crime – and to consider the *distribution* of each characteristic in a particular group.

In 1840 he turned to the subject of the distribution of specific characteristics in a population and his work immediately bore fruit. What inspired Quetelet to put his data into diagrammatic form is not as important as the fact that he did it. He divided the different heights that were possible in a group of people along an horizontal axis, and then noted the total numbers of people of specific heights in columns parallel to the vertical axis.

Now much of science progresses by trying to find some new connection of one variable with another, and in Quetelet's doodling with diagrams we find a typical example of the leap of insight from casual notion to startling revelation.

What struck Quetelet so forcibly was the shape of the diagrams his data led him to draw. The highest columns in his diagrams clustered around a mid-point in the range of possible values of the height variable, with the highest column in the middle of the range.

To envisage the shape of Quetelet's diagrams, assume that the majority of adult males are about 5 feet 8 inches in height. It follows that if we graph (see figure 7.1) the numbers of males at each particular height, say, in intervals of one inch, the tallest column would be the one representing the most common

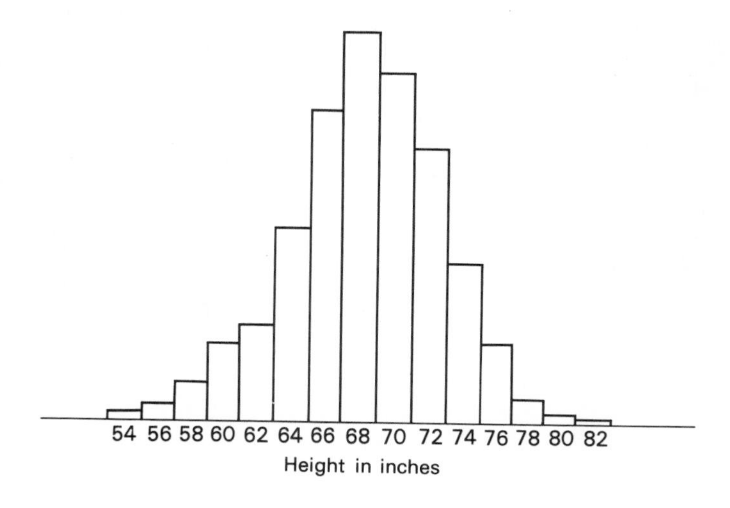

Figure 7.1 Distribution of males by height

height and the smaller columns the less common heights. There would be some males over 6 feet 8 inches, though not many, and some would be under 4 feet 8 inches, again not many. In between these heights the other males would be distributed, with the majority clustering around the most common height. For women, the distribution would likely cluster around a mean height of about 5 feet 4 inches (Quetelet, of course, worked in metric sizes, but this would not affect the shape of the distribution).

The height of the columns in Quetelet's diagrams fell away symmetrically on either side of the highest one until, at the extreme values of the range, the columns were very small – because very few people were as tall or small as the extremes of height that are possible for the human frame. The majority of the population clustered around the middle of the height range.

Recalling the distribution of errors in astronomical observations from his Paris days, Quetelet was struck by the similarity in the symmetrical shapes he drew from his data on the variation in the human form and the symmetry of the error curves of the astronomers. The majority of astronomical observations tended to centre themselves within the range of recorded measurements, with a minority of observations, representing greater errors, spread symmetrically around the mean.

For the scientist, such coincidences are the very El Dorado of a life's work – fortunate is the scientist who finds one that really matters (all that glitters is not gold). Quetelet's discovery was one of those that really mattered.

He went on to state the view that the human form deviates from a norm or average according the some law of chance. The multifarious influences that go to make up a particular form, interact to produce a shape that deviates from an average or standard shape by a little or a lot depending on the exact mixture of the influences, but the tendency is on average for the influences to produce shapes and sizes not much different from the norm or ideal shape or size.

This distribution is much like the shots at a target by a rifle. There might be some stray shots, wide of the bull's eye, but the majority cluster round the bull. What caused a particular shot to land where it did could be thought of as the result of many influences on the shot from the time the trigger was squeezed to the moment it hit the target.

The wind, the exact amount of propellant in the bullet, the pressure of the trigger finger and so on, all contribute to the trajectory of the bullet on its way to its impact. Sometimes a particular influence – the wind – would work one way while the trigger pressure would work another, and a third or fourth variable would work contrary to those experienced in previous shots. The result was a distribution of the shots around the bull's eye.

Similarly in nature. The multifarious characteristics that influence the specific shape and size of the human body operate to produce something deviant from the most common form. The majority of deviations cluster around this ideal form, and the wider the deviation the fewer the examples of it that are found.

Apparently, nature had a greater control of the chance events that influence outcomes than was realized – it strove for perfection and conformity and approached it in the majority of cases. The artist in Quetelet agreed that *l'homme moyen* (or average man) was beautiful, as nature intended, but chance interactions on various aspects of the human form combine to produce something less than perfect in one degree or another.

Quetelet's revelation appeared to some to snatch mankind back from the chance events of a random world, into an orderly God-divined universe. Change and deviation were constrained and predictable if a sufficient number of incidents were considered. The chances of wide deviation in any particular characteristic were limited, and the majority of people conformed to some common average characteristic.

Moreover, the distribution of a characteristic in a population when put into a diagram, conformed to a specific shape. This can best be described literally as 'bell shaped'. To remind you what a bell shape distribution looks like consult figure 7.2 and note how it is symmetrical around its vertical axis.

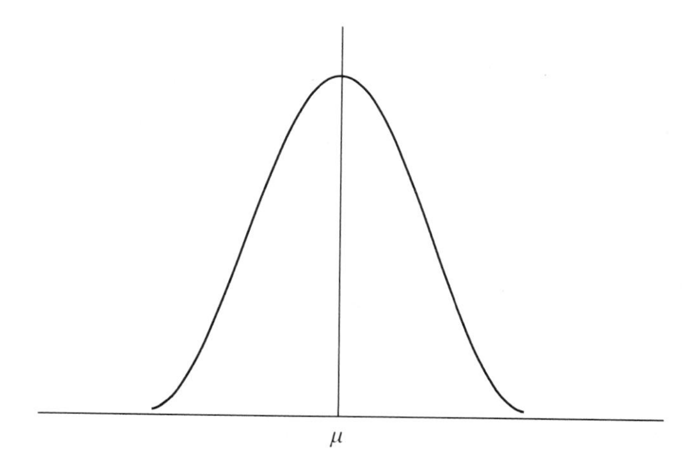

Figure 7.2 Bell-shaped normal distribution

The properties of such a bell shaped distribution constitute the essence of modern statistics. Study that bell shape, understand its properties, and you have elementary statistics, and much of the work up to advanced level, in your grasp.

8

Chance is a Fine Thing

The man who consorted with gamblers

In the early years of the eighteenth century, Slaughter's Coffee House in Long Acre, London, bustled with what less wordly authors call the 'seamy side of life', by which they mean, in part, that the place was peopled by gamblers and their then kinsmen, annuity brokers. This was no diamond-chandliered green baized plush casino where the rich and famous frittered away their time (and fortunes). It was a grubby place, noisy and vulgar, much closer to a modern betting shop in a depressed area, with perhaps a touch of the kind of slot-machine 'palace' that is common in London's Soho.

In this den of moderate iniquity, a gentleman with a foreign name, and a somewhat short temper, sat at a table and idled away his day while waiting for clients to purchase his professional services. His name was Abraham de Moivre (1667–1754). He was a protestant refugee from catholic France, and also an accomplished mathematician.

De Moivre's special talents lay in the newly emerging field of statisticial probabilities. He could calculate, to order, the odds for gamblers' bets in games of chance, be they dice or cards, or the annuity rates for those who gambled for or against the likely life-spans of those with money and a desire for easy income.

De Moivre was no unknown or unsung genius. He was the friend and colleague of many of the brightest mathematicians of his day. He is described as being 'intimate' with Isaac Newton, and he was in his own right a member of the Royal Society. De Moivre's problem was that he was also poor; too poor to afford a marriage, and not quite rich enough to afford much else. Hence, he undertook the entirely honourable course of working for his living, using the innate skills he was blessed with.

Karl Pearson (himself a brilliant statistician) wondered about De Moivre's need to work in this way, and in a rare glimpse of his own imagination, pictured the pathos of the genius at work among those of less (and totally forgotten) ill-repute:

> I picture De Moivre working at a dirty table in the coffee house with a broken-down gambler beside him and Isaac Newton walking through the crowd to his corner to fetch out his friend. It would make a great picture for an inspired artist (Pearson, 1978, p. 143).

De Moivre may or may not have deserved our sympathy. Perhaps he enjoyed his work too! Not all frequenters of the baser side of society are in a perpetual state of misery. Far from it. And we have no reason to suppose that De Moivre cringed each morning as he set out to Slaughter's Coffee House, weighed down with debt and worry. Perhaps he did not believe that the coffee house tore him away from his scientific interests.

For all we know, De Moivre set off each day with a spring in his step, anxious to get to his 'laboratory' to run yet more field 'tests' of his probabilistic theories! Each gambler ('broken down' or otherwise) was for De Moivre a new case for his book, *The Doctrine of Chances,* (1718), and each annuity broker (were they 'broken-down' too?) provided yet more evidence for his other great work, *Annuities Upon Lives* (1725).

Both books went into several editions, and there is no doubt (in my mind at least) that the frequent call to exercise his talents in the coffee house tested and re-tested his theories in a way that would not have been improved if De Moivre had been living high and well, complete with a large research grant from the rich and famous. Moreover, perhaps De Moivre enjoyed gambling too!

Whatever the truth, the fact remains that de Moivre was the first to outline the mathematics of what we know today as the normal curve. Recognition came late to de Moivre. Long after he had passed on, the world credited the normal curve first to

Gauss (it become known as the 'Gaussian' curve) and then to Laplace and Gauss. De Moivre was unlucky, for another of his discoveries – the limit of a binomial distribution – was credited to Simeon Poisson (1781–1840), and it is known today as the Poisson Distribution!

It was Karl Pearson who noted de Moivre's statement of the properties of the normal curve in his researches (1924) while preparing his famous lecture series on the history of statistics, and, slowly, the credit for discovering the curve and its properties has been passed to the man who preceded Laplace by forty years.

What is a normal distribution?

You have already been introduced in chapter 7 to the concepts of the mean and the standard deviation and we want now to look at why statisticians consider these concepts of importance in relation to the normal distribution.

Let me begin by saying that the mean and the standard deviation of a population are as important to the shape of the normal distribution as the centre of a circle and its radius are important to the shape of a circle. Define these two properties and you have a circle; define a different radius and you have another circle. Of course, all circles have the same shape geometrically; no circle can have different radii.

In the case of a normal distribution, the mean and the standard deviation together define its shape. For different standard deviations there are different shapes, though all shapes of normal distributions have a common property: the curve of the distribution is symmetrical about the vertical axis, that is, there are as many items in a population deviant in value from the mean to the right of the axis as there are to left of it (see figure 8.1).

You will recollect from chapter 7 that the standard deviation is calculated by finding the square root of the sum of the squares of the deviations of each item in a population from the arithmetic mean of the population. The size of the standard deviation relative to the mean tells you how dispersed the items in the population are from the average for the sample. The larger

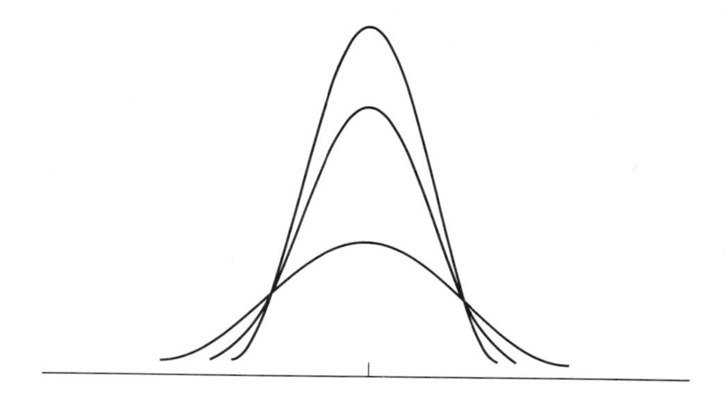

Figure 8.1 Normal distributions with different standard deviations

the standard deviation relative to the mean the more widely dispersed are the values of the items in the population.

The standard deviation calculated in this way has an interesting property in respect of the normal curve. The shape of the normal curve is such that if you add the standard deviation to the mean and mark off this point on the horizontal axis, and then take the standard deviation away from the mean and mark this point off on the horizontal axis, you will find that these points are vertically below the the points on the normal curve where the shape changes (where, in technical language – explained below – the shape changes from being concave downwards to being concave upwards).

In case this imagery blinds you, look at the normal curve in figure 8.2 and note how the curve drops from its highest point (where the mean of the population is represented) at first steeply, relative to the axis, and then drops less steeply, relative to the axis, as it approaches the horizontal.

The points on the curve where this change in direction occurs from steepness to less steepness are known as 'points of inflexion'. In the normal curve these turning points are vertically above the points representing the mean plus or minus the standard deviation.

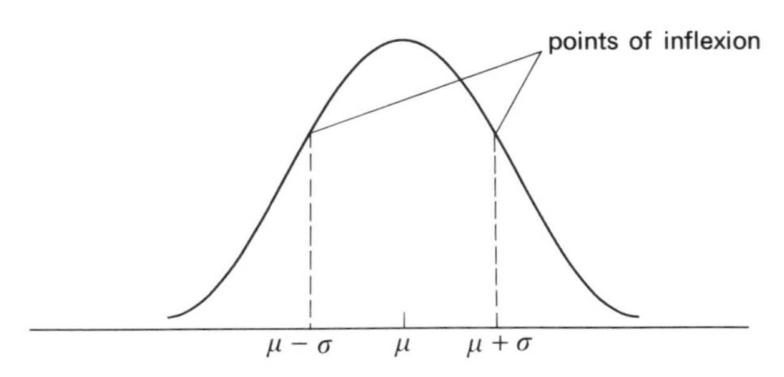

Figure 8.2 *Standard deviation and the normal curve*

Chance and the normal distribution

In Figure 8.2 you will find that the mean of a normal distribution is given the Greek letter mu (μ) and the standard deviation the letter sigma (σ). Hence, we mark off points, which are plus or minus the standard deviation from the mean, as mu ± sigma ($\mu \pm \sigma$).

So what? Well, for a start, the normal curve already has a property (obvious to you once it is drawn to your attention) upon which you can develop some powerful ideas about the distribution of items in any population *that is normally distributed.*

If the items in a population are symmetrically distributed above and below the mean value, it follows that 50 per cent of the population have values greater than the mean, and 50 per cent less than mean. Put this another way: what are the chances that a item is above the mean in value? Why 50-50, of course! Can you begin to see a glimmer of the possible use of such a normal distribution?

It does not matter whether the population is distributed symmetrically about the mean in such as way as to have the majority of its values very close to the mean (thus, producing a very 'tall' normal curve) or widely dispersed about the mean (thus, producing a very 'shallow' normal curve) (see figure 8.1), we know that half of the items are greater than the mean and half of them less than the mean. The area under the curve can be considered to be squeezed or compressed, depending on the characteristics of the population in question, but every normal curve has an area under it representing a single unit area (that is, all the values of the population). If we conceive of this area as being unity, the vertical to the mid-point of the normal curve divides that area into two equal portions.

Holding this idea in mind, let us return to the relationship between the mean value of the curve and the standard deviation. Is there a specific relationship between the items that fall within the range of one standard deviation above the mean and one standard deviation below the mean? Yes there is! I have already asserted that the points representing plus or minus the standard deviation about the mean value are immediately under the points of inflexion (where the curve changes from steep to less steep). Such is the mathematical property of the normal curve – a testimony to the genius of De Moivre – that we can state categorically the exact proportion of the population that will lie between any two values of items in a population (figure 8.3).

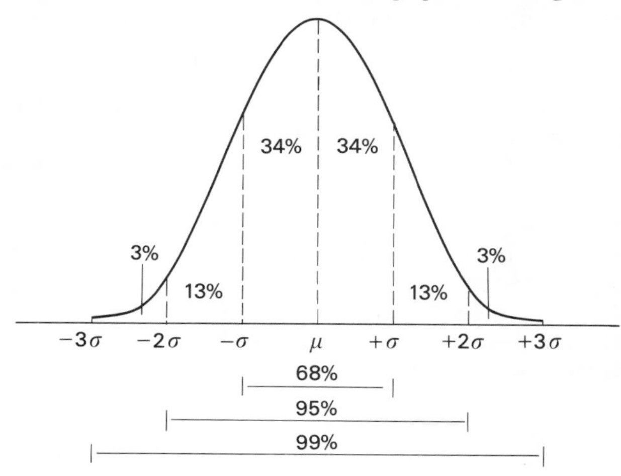

Figure 8.3 Approximate areas under normal distribution

We are not confined to the relatively uninteresting assertion that half the population has values above the mean and half values below the mean because we can go much further and state exactly what proportion of the population will have values between, say, one standard deviation above and one standard deviation below below the mean, or, between any given value (expressed in standard deviations or multiples thereof) above the mean and another value above (or below, or equal to) the mean.

For example, around 1/3 (actually 34 per cent) of a normally distributed population will be found within the range of the mean plus one standard deviation, and another 1/3 within the range of one standard deviation less than the mean (this follows from the symmetry of the distribution). Adding these two together we can assert that about 2/3 of a population lies within plus or minus a standard deviation and the mean.

Consider this in probability language. Statisticians assert (and can prove mathematically) that 68 per cent of a population will be found within plus or minus the standard deviation of a population's mean value, or, putting it another way, if we are asked what are the chances that a member of a population is no greater or no smaller than the mean value plus or minus the standard deviation of the population, we would be able to answer immediately and precisely.

The value of being able to do this is clearly high. You can look at the data and instead of seeing a mere mass of numbers you can organize it in a form that provides you with a means of predicting probabilities that particular values of the data are found within discrete ranges from the mean value of the population.

Let me illustrate this point. In our example here we want to know the chances of a member of a population being within a much narrower range (plus or minus a standard deviation) than the entire range of the distribution (which stretches several standard deviations either side of the mean).

Plus or minus one deviation from the mean covers 68 per cent of the population and therefore the chances of an item falling within that range is precisely 68 per cent, or, in the language of probability, 0.68 (where probability is measured by a number between 0 and 1, and where 1 indicates absolute certainty).

It follows that if 68 per cent of the sample is inside the range of plus or minus a standard deviation about the mean, then 100 –

68 = 32 per cent must fall outside this range, that is, have values greater or lesser than plus or minus one standard deviation about the mean. Moreover, as the normal distribution is symmetrical, we know that 32/2 = 16 per cent of a population will have values greater than the mean *plus* the standard deviation, and 16 per cent will have values below the mean *minus* the standard deviation.

Suppose, by way of illustration, we have decided to pass all students who score 50 marks or more in an examination and suppose in addition that the results are normally distributed. If we found that the average mark achieved by the students in the examination was 60 and the standard deviation was 10, what proportion of the student body would we pass?

We know that half the students must have marks more than the mean of 60 because half of a normally distributed population exceeds the mean value, hence we know our answer is at least more than 50 per cent of the students have passed the examination.

The standard deviation of 10 taken from the mean equals 50 (which happens to be our pass mark) and we know from our assertions above that 34 per cent of the population will have marks between the mean minus one standard deviation and the mean, and hence we can confidently assert that 34 per cent of the students got marks between 50 and the mean of 60. Adding the two proportions together: 50 per cent above the mean of 60 plus 34 per cent between 50 and 60 gives us 84 per cent as the proportion of students who have passed our examination.

From this idea we move on to another. If one standard deviation above or below the mean covers 68 per cent of the population, what proportions of the population fall in ranges beyond a standard deviation either side of the mean?

All these proportions have been worked out by mathematicians (and are available in statistical tables for you to look up) and this should give you an insight into the properties of the normal curve.

I will assert here, but not prove, that we can take unit distances, measured in multiples of the standard deviation, along the horizontal from the mean in either direction and predict the proportion of the population that would fall within the mean value and that multiple of the standard deviation. What in effect we are doing is taking the area under the curve between two points on the horizontal as a proportion of the total area of a normal curve, and converting these proportions into measures of probability. For instance, if you took a range of approximately two standard deviations (the standard deviation of the population times two) above and below the mean you would find 95 per cent of the items in the population would have values between these extremities.

Taking our examination results, for instance, where the standard deviation is 10, we can assert that approximately 95 per cent of the students would have a mark between 40 and 80 (because $60-(10 \times 2) = 40$, and $60 + (10 \times 2) = 80$). The remaining 5 per cent of the students would have marks either above 80 or below 40.

Again, using probability language, the chances of a student having a mark between 40 and 80 is approximately 0.95. I have said 'approximately 95 per cent', because the actual range for this proportion is slightly less than two standard deviations (1.96 times the standard deviation to be precise).

You can note that the range of approximately three standard deviations above and below the mean covers 99.7 per cent of the population. For most of the work that you will be concerned with in social science, approximately 95 per cent of the values in the populations you are interested in will lie within two standard deviations of the mean. These simple properties of the normal curve will eventually become your everyday tools of the trade, so to speak, and you will be able to make all kinds of judgements about your data (assuming that it is distributed normally – there are other distributions in statistics, including skewed distributions, but for now you should concentrate on the normal curve only).

The Z score

You are now in a position to understand one of the many tests that can be applied to your data to draw out its comparative significance. It uses nothing more than the information that you already have to hand in the earlier discussion on the properties of the normal distribution.

Let us consider the familiar case of a coffee room discussion between students who are informing each other of their examination results. Typically, student friends may have sat different groups of subjects, some having taken economics, law and marketing, others economics, administration and politics, or sociology, law and history and so on.

It is always impressive to be able to announce results in the high 60s or 70s and less so to have to confess to bare passes in the low 50s. You would be a most unusual person if you were not chuffed at getting a 'good' set of examination marks, and, whatever your public disclaimers about your merits, it being vulgar to gloat, you would be less than normal if you disdained to glow a little inside at your success (and feel not a little dispirited by any failures).

But can you really draw any conclusions from a good mark in the 70s in, say, marketing compared to a mark in the 50s in, say, politics? In so far as you might make judgements about your course choices for later years on the basis of your results – tending to associate 'good' marks in a first year with your prospects for good marks in subsequent years – you could make a disastrous choice of subject which you will regret once the real pressure piles on.

For a variety of reasons, some courses in some years can have relatively easy examinations which it is not difficult to score well in, and as a direct result – for the staff look at the results too! – in the following year, the examinations are made more searching of the students' abilities. You could score 72 in a first year class and be struggling to get 50 marks in the second year class.

When I was a student counsellor I spent a not inconsiderable amount of time gently making students aware that the pass marks they get are not scored against an absolute standard common to all examinations in all departments: 70 marks out of 100 does not have the same absolute meaning in each examination you sit. A little knowledge of statistics could have saved more than one student from making a course decision that they came to regret in later years. How can you avoid being misled by your apparent success (or failure)? Simple: consult the Z score.

The Z, or standard, score is a means by which we compare

different scores across different populations. We are, in effect, turning disparate measurements into a common measurement. We do this when we measure a particular length by stating it in terms of a standard measure of distance - be it inches or centimetres - and when we measure a weight against some common standard. In statistics we do not have a standard measure of the standard deviation of a distribution, but we can compare scores by comparing values in the distribution with the standard deviation of the distribution. For this we have to do a little arithmetic.

It is not the mark you got out of 100 in your examinations that is decisive. That is only half the story (though this is adequate enough for your grandparents and favourite aunts, if you want to impress them - and why shouldn't you?). To compare two marks you need to know the mean mark of the group who sat the examination with you and the standard deviation of the marks for that group. Put simply, consider your mark of 72 in a marketing examination where the mean mark was 59 and the standard deviation was 10. Immediately, from your knowledge of the properties of the normal distribution, you know that 68 per cent of the class scored between 49 and 69 (because 59 plus or minus 10 gives the range 49 to 69), and that only 16 per cent of the class did better than 69. You are in the top 16 per cent of the class, surely a good result?

If, however, the mean score in marketing was 53 and the standard deviation was 5, then you know that 68 per cent of the class scored between 48 and 58 (because $53 + 5 = 58$; $53 - 5 = 48$). This means that 16 per cent did better than 58 (and you were one of them!) and 16 per cent did worse than 48.

Compared to your mark of 72 the mean mark of 53 seems a long way off from your result. Indeed it is. You have done better than three times the standard deviation above the mean ($3 \times 5 = 15$ and $53 + 15 = 68$ which is less than your 72) and you know from the normal distribution that 95 per cent of the group will have scores of under three times the standard deviation above the mean. You are not only in the top 16 per cent of the class, you are also in the top 5 per cent, which is cause enough for you to feel a trifle smug.

The above might suggest to you a way of comparing one mark

with another. Instead of comparing the absolute marks, we should compare their relationship in multiples or part multiples of the standard deviation to the mean.

To have a mark that is 2.5 standard deviations above the mean is much better than one that is only 1.9 standard deviations above the mean, irrespective of the absolute values of the mark. Conversely a mark that is within 1 standard deviation below the mean is a better mark than one that is 2 deviations below the mean, again irrespective of their absolute values.

Consider the case where you have scored 72 marks in marketing and somebody else has scored 63 in politics. Which of you did best? That depends entirely on the mean of each class and their standard deviations.

If in your marketing examination the mean mark is 68 and the standard deviation is 10, while your friend's politics examination produced a mean mark of 52 with a standard deviation of 3, there is no doubt that your high 72 is not as good a performance as your friend's 65. Why? Because your 72 is *less* than one standard deviation above the marketing class average ($68 + 10 = 78$), while she is more than three standard deviations above the politics class average ($52 + 3 + 3 + 3 = 61$).

If on the basis of your pass of 72 you rushed into Marketing II, you could end up selling brushes on the doorstep instead of masterminding a marketing campaign for brush sales staff, while your friend, if she as rashly avoided Politics II, could miss her chance of ending up as Prime Minister.

Calculating the Z score is slightly more precise than I have illustrated here. However, the arithmetic is simple: just take the class average mark from your own mark and divide it by the standard deviation. This will produce a number in as many decimal places as you need for exact comparisons. In the cases mentioned above your exact Z score is $(72 - 68)/10 = 0.4$, and your friend's is $(63 - 52)/3 = 11/3 = 3.6$. Clearly, 3.6 is bigger than 0.4 and her achievement is more impressive.

Nature is niggardly

Data does not get handed to a scientist in neat packages, all carefully worked through and complete in every respect.

Nature, as Professor Robbins pointed out, is niggardly. It gives up its data reluctantly. The scientist often has to go to extreme lengths to acquire data and, when it is ready for examination, its incompleteness is its limitation.

Also, in collecting data of a particular phenomenon the scientist has to be selective, and in being so it is necessary to leave out much that may in fact be an important, though hidden, influence on the phenomenon that is being investigated. For example, you might believe that you are collecting data on the proclivity of certain social groups to deviance (a less prejudiced word than criminality) but your data may come from areas totally unrepresentative of the population as a whole and unrepresentative of similar social groups in other parts of the country. Your data could lead you to pick up what appears to be a rising trend of deviance across the country compared to previous years, and yet if, for example, you were aware of the relative number of laws operating in one period compared to another, you might consider the evidence open to another interpretation.

A simple thing like the necessary selection of data from the mass of information available could bias the trend. This was certainly the case with the the 'crime wave' of the late eighteenth century – as more laws were passed, more people ran foul of them; this encouraged the authorities to impose ever more draconian punishments (hanging and transportation), and, as this apparently did not stem the numbers convicted in the courts, it provided more 'evidence' of the 'crime wave' sweeping the country provoking more laws and yet more repression. The result was widespread anxiety about a breakdown in 'law and order' (which the French Revolution did little to calm) provoking greater determination in local magistrates to apply the 'law' and in the gentry to suggest new ones. In this way a study of selective data – the number of people convicted of crimes – that ignores the wider social context could mislead you into quite spurious conclusions.

Statistical techniques are aimed at minimizing the opportunities of falling into, largely avoidable, errors of this sort. Much of statistics is concerned with establishing procedures to check the probity of selected data and to set minimal standards for designing methods of data collection and interpretation that will pass muster within the particular discipline to which you belong.

As with a jury deliberating on the guilt or innocence of a defendant, so the members of a discipline make judgements on the value of the work of their peers. They are guided in this by the extent to which their work meets established criteria of evidence. Like some juries, the profession can get it wrong; sound work can be, and is, dismissed for all kinds of reasons, not the least of which is the possibility that it challenges established conventions or, worse in the eyes of some 'scholars', it offends the authorities. Unlike a jury, however, there are not just two verdicts of guilty or not guilty (three in Scotland where the verdict 'not proven' is time-honoured). Your work is 'marked' along a continuum running from 'probably correct' to 'almost certainly not correct'.

But there is nothing like a fundamental error in evidential procedures for a piece of work to be dismissed out of hand. Of course, there is also much debate on what constitutes relevant evidence for or against a hypothesis – as a glance at the journals of all disciplines will show – but nothing kills a hypothesis faster than an absurdity in the use of data.

It is in this area that statistics has most to contribute because it provides the scientist with a means to identify the available data and to decide within reasonable and consistent boundaries whether the data is representative of the phenomenon being investigated or not. In doing so, statistics sets out in a rigorous language the reasons why the evidence has been judged in this way.

9

By Your Samples Be Judged

What constitutes sound experimental practice?

The techniques of statistics are international; it does not matter in which language the scientist works. If the data is manipulated to specific and valid criteria it is possible for a scientific inference to be made from it and, if the data is properly presented, it is possible for other scientists working in different languages and in different continents (or even centuries) to check the probity of the work and to comment sensibly on the conclusions.

It is not at all obvious that a particular experiment provides sufficient evidence to draw a valid conclusion. In daily life we meet examples of firm conclusions being drawn from a sample of one (usually our own limited experience). You must have heard somebody argue along the lines that there is no relationship between smoking and cancer because 'my aunt smoked sixty a day and lived till she was 90, while my uncle never smoked and died of cancer at 65'.

In these discussions it is wise to correct the validity of such 'evidence' only once, because if it continues to be asserted by that person they are obviously immune to scientific reason, and you will obviously upset them and yourself needlessly if you keep dismissing the use of a limited sample – as Schiller put it: 'against stupidity the very Gods Themselves battle in vain'.

What constitutes reliable evidence is often hotly debated among social scientists (look for example at the debate on the evidence for and against a connection between intelligence and race) but, whatever the continuing controversies, they are at least conducted in a broadly common approach to scientific purpose, and I am putting it to you here that in this common approach statistics can be seen as a set of tools for the collection and assessment of evidence.

Francis Bacon made some attempts to formalize a method for conducting experiments. He experimented with wheat seeds soaked in different fluids (wine, urine and water from cow dung) and compared their germination rates to wheat seeds that were not treated at all. The result of Bacon's experiment (he found urine the best treatment!) are less important than the method he was using: he compared treated seeds with each other *and* with non-treated seeds. The non-treated seeds acted as a *control* group for the experiment, and the idea of using a control group is now a feature of the experimental method mandatory in the social sciences.

Arthur Young, the eighteenth-century English agricultural experimenter and innovator, made extensive use of an experimental method which stressed the need for strict comparability between samples and also emphasized the benefits of obtaining as many results as were feasible. He realized that for authoritative conclusions the utmost efforts at rigour were essential and that the authority of an experiment would increase as the number of examples increased.

Experimental skill requires careful preparation and strict attention to detail. For Young, this was an essential feature of his work. He tried to ensure that the conditions of each of his seed experiments were the same, or as near the same as was feasible, in each experimental plot. He did not trust the results of an experiment in a single year and insisted on repeating them over several years with different arrangements of the plots.

One thing he steadfastly refused to do was to interpolate results either from his previous experience or from non-comparable tests. 'Nothing of this sort should be done' he wrote, 'even on the same soil, unless the experiments were absolutely comparative; it may be a matter of amusement or curiosity, but of no utility, no authority'. This is the true voice of science and one that is echoed by all who would make scientific judgements about anything remotely connected with evidence.

It took some considerable time for the conditions for sound

experimentation to become the norm rather than the exception in science (both social and natural) and for the tools for a sound quantitative interpretation of the data to be developed. Much of the work of statisticians up to the Second World War was about developing sound statistical techniques for coping with data and making inferences from it.

From experiments to social statistics

An interesting aspect of the history of statistics has been its affinity to various social 'movements', particularly in the nineteenth century when 'improvement' and 'progress' were seen either to be necessary or to be inevitable for the human race. This produced a curious mixture of reformers: those trying to improve the world by calling for action to stem the breeding of the 'least suitable' specimens in society, in the belief that if this policy was followed it would improve mankind genetically, and those other reformers (though sometimes they overlapped in personnel) trying to improve society by nursing the casualties of social change, that is, the very poor. Among the latter we could include the work of Le Play in Paris, Patrick Geddes in Edinburgh, Charles Booth in London and the general efforts associated with the early Fabians and the Labour Party, and the Liberal governments, to introduce the elements of a welfare state.

In some situations the reformers sought data for confirmation of their views; in others they sought techniques to handle their data. Statistics developed through this period with contributions from both approaches. The motivation to do so was provided by their almost evangelical views on what they believed was good for society.

We can see an example of motivated research in the *Eugenics* movement of the years up to the First World War – eugenicists believed that human improvement was possible if hereditary laws were applied by a careful policy of preventing some people from breeding at all, such as the infirm, the stupid, the alcoholic and the hereditarily diseased. The object was to raise the standards of the race in health, intelligence and social behaviour. There were strong affinities between these ideas and those of social Darwinists ('natural selection of the fittest' and so on).

Eugenics attracted the leading statisticians of the time: Francis Galton (1822–1911), Karl Pearson (1857–1936) and R. A. Fisher (1890–1962) were leading eugenicists and they produced most of the excellent work in statistical techniques that has lasted (you will find their creative work abounds in statistics, though you will be hard put to find modern references to eugenics; it died out as a movement in the 1920s).

It was in the field of biology that the inspiration for developing statistical techniques was most prevalent. Darwin's theory of natural selection and his experiments to establish it promoted interest in the measurement of differences between similar plants in different environments. Mendel's work on genetics gave this speculative work an impetus too, and attention turned to establishing the extent of the genetical influences on organisms.

Scientists like Galton set to work, combining a concern for accuracy in the experimental role with a search for adequate techniques of establishing relationships between one thing and another. This led Galton to the techniques of correlation and regression (discussed later in chapter 11).

Other scientists concentrated on designing experimental conditions peculiar to the needs of biological research, particularly in agriculture. Different varieties of seeds of the same plant had to be tested in broadly similar conditions if anything worthwhile was to be said about their relative performances.

It was not sufficient to try one type of seed in one field and another in the next field because there could be serious differences in the soil, moisture content and such like, operating on the seeds to produce differences in growth that were not obvious to the experimenter. These could seriously damage the validity of the experimental conclusions for circumstances where these conditions were absent.

If variety A appeared to grow faster and have a better yield than variety B it might have more to do with the fact that the field planted with A had better access to the sun and was more sheltered from the wind than the field with the B variety in it. It would be useless to recommend using variety A if this was the case, if only because it would waste the farmers' money and, as a result, discredit the advice of the researchers in future (bearing in mind how research grants, and therefore researchers' incomes, are often, though not always, linked with the credibility of their research). One variety of seed, with no real differences between it and other seeds, might grow faster because the soil fertility was more favourable in the field chosen for it to grow in.

This problem caused much literature to be written on how a field could be sowed in small plots in such a way that the possible incidence of soil or environmental differences could be eliminated from the experiment by spreading the possible differences in soil and other conditions across the entire experiment. If the hidden influences were neutralized by plot design, then any differences in performance that were captured by the statisticians could reasonably be attributed to differences in seed variety alone.

Experimental designs associated with crop productivities required some means of combining the different results from each variety for comparison with the different results of other varieties, assuming that the pattern of sowing the varieties eliminates or reduces to insignificance the potential external influences of the environment where the experiment happens to be conducted.

On the experimental design side, the sowing of alternate rows in the same field was one obvious method of reducing the influence of field fertilities but it was not altogether satisfactory because soil conditions could vary between the strips. Many combinations of patterns were advanced and tried out. In principle, sowing in small plots according to a predetermined pattern across a single field was an obvious solution. The questions that arose were: which pattern, and how big were the plots to be in how big a field?

Scientists wanted to discriminate accurately between seed varieties and their yields, and the experiment had to be open to replication by others to get similar results for the varieties. It was decided that different varieties of seed must be tested in plots next to each other and that any inherent differences in soil fertility had to be eliminated horizontally, vertically and diagonally. The plot patterns produced a number of ingenious designs (my own 'favourite' is the plot pattern that moves individual plots around much as the knight piece moves in chess).

Though this work was concerned with field experiments in agriculture, the fact that statisticians recognized that there was a need to eliminate the chance of hidden variations in those factors that could influence the outcome, which is implicit in the debate on the 'proper' lay-out of a field experiment, helped considerably to make sound experimental procedures an explicit principle of statistical work, and this has carried over into the social sciences: hence, what seems at first sight to be a pedantic approach to research standards in the social sciences.

You will soon become aware of these issues as you read your journal articles and research reports (and the debates about them), and you should note that this is not just a question of doing our arithmetic accurately – important as that is of course – nor is it a sign of a 'too fussy by half' approach by people with little to do but make studying difficult for students!

Sampling methods

High standards in the design of a research method are most important aspects of the structure of any research programme that is attempting to answer important questions about the world we live in, and high standards are necessary to ensure there is some relationship between the problems we address as social scientists and the answers we generate from the data (see Spector, 1981).

If an experiment must be designed properly in agricultural research, to take account of the possible variations in soil and other conditions, it follows that in the case of studying variations in human behaviours we must also design things so as to eliminate possible variations in the environment that can influence our conclusions, if only because the variability of human behaviour is much wider than with natural phenomena.

For example, in political science if we are studying political attitudes, say, and want to know how the electorate at large feel about some proposal or other, it would not help the validity of our conclusions if we only asked people in the vicinity of a political party's annual conference near lunch-time, as it is likely that the views of that particular party would be excessively represented in our survey, especially if we interviewed 'passers by' outside the restaurant nearest to the conference hall!

In one famous example in the 1948 US Presidential Election, an opinion pollster got the wrong result because he polled voters by telephone which gave him a higher proportion of Republican voters (who were in the main richer and more likely to own a telephone than voters who supported the Democrats). He predicted a landslide for the Republican candidate, Dewey, over his Democratic rival, President Truman.

This poll is now put up for semi-mockery in politics and market research classes, yet in those early (for social science) days the people concerned were vulnerable to the howlers they committed largely because they were not aware of the work in experimental design in the biological sciences carried out 50 years earlier. This deficiency has now been corrected and the rigour of the research procedures of reputable polling organizations (whether studying political preferences for different candidates or market preferences for different brands of consumer goods) are comparable with the best standards found in laboratory research.

Let me develop this point by reference first to the question of samples of data from large populations. The scientist, in the absence of being able to examine the entire population, has to make do with smaller quantities of data than would be available if the examination of the entire population was not excluded on the practical grounds of cost, time and access. Nobody, for instance, lives long enough to study the entire universe, hence we take samples of it. A sample is any subset of a population.

We can approach the sampling problem from two ends: either we can say that we are bound to work with samples rather than the entire population because of the impracticalities of studying the entire phenomenon in sufficient detail, and, therefore, we need to know if our sample is representative of that population (that is, is the sample from the population at all?), or we can ask the equally interesting question of how small or large a proportion of a population is it *necessary* for us to study in order to make valid statements about it?

Important questions are raised about the minimum size of a sample sufficient to be representative of the population from which it is drawn. If the sample is too small (in the absolute sense, for example, under 30) it might lead to erroneous conclusions about the nature of the population; if it is too large (more than 10 per cent), you might be wasting resources which you could use elsewhere.

In both approaches to the sample-population question we have a commonality of interest. We are seeking the minimum acceptable *degree of confidence* that a particular sample does come from the population we are interested in and not some other population, and that it is truly representative of it.

In the one case we want to be reasonably sure (nothing is certain) that the data is connected with the phenomenon and in the other we want to know beforehand just how much (or little) data to collect to reach the minimum amount necessary to provide results that we are reasonably sure reflect the phenomenon concerned.

Applications of this approach spring readily to mind in the fields of quality control – how many light bulbs do we test out of a batch to be reasonably sure that the batch is of a certain standard? – and in political polling – how many voters must we interview, *and how do we select them,* to be reasonably sure that the declared voting intentions correspond to how the electorate as a whole would actually vote?

This is not to say that there is only one correct method of conducting a poll of views or a test for quality in a batch of output. It all depends on the circumstances, and there is a variety of polling methods available to take account of different purposes and different degrees of generality.

In some cases we are looking at specific social groups and searching for information about their unique behaviours; in others, we are looking for general information representative of the widest possible population. In all cases we want to ensure that the selection of our sample is as random as we can get it because we do not want the researcher's explicit or implicit biases to creep into its selection (she only interviews good-looking men, he only interviews whites).

We must look out for sampling errors that could influence our results, such as could arise from ignoring heterogeneity in the population. This is a factor that could be decisive, such as making sure our sample covers all tax-paying income groups if we want to find out the attitude of the population as a whole to taxation.

In some cases the heterogeneity of a population may not matter in respect of what we are looking at and we can safely ignore it. This might be the case with road accidents or the number of people requiring blood transfusions over the Christmas holidays.

If a population has identifiable characteristics that could be important in our sampling of opinions or behaviours, we must ensure that the people interviewed or observed appear in the sample we study in the same proportions as they are represented in the population. Stratified sampling is one way of achieving this and much care is taken to ensure that the sample is as representative of the population as can be.

In researching the views of students on grants versus loans as a means of funding higher education we ought to ensure in our sample that the proportions interviewed reflect the male–female balance, just in case there is a difference of view between these distinct categories. Similarly with the balance between engineering and sociology students, undergraduate and post-graduates, under 20s and over 40s and so on.

It might also be advisable to ensure that our sample truly reflected the geographic spread of the country if we intend to generalize from our results. Students in Scotland might feel differently about loans than students in Cambridge; *a priori* we do not know if they do and we ought therefore to ensure that we include both groups in our sample in case we generalize from an incorrectly defined sample.

Moreover, you can bet your research grant that if you ignore these fairly obvious precautions then those who comment on your work professionally – usually your peers and your emloyers, and anybody who does not agree with your findings – will search for 'flaws' in your selection of data, and among the most obvious charges of error that you will invite are those associated with sample selection.

There is something devastating about a comment such as:

'Very interesting findings but can we be sure that the views expressed by some engineering students in the beer bar at Strathclyde University are really representative of what all students feel about the issue elsewhere in the country?' If your sample is poorly selected you cannot honestly answer that kind of comment and if you cannot answer it honestly the implied criticism, justified or otherwise, will stand. It is best to anticipate the question by ensuring your research methods do not give rise to it in the first place.

Here we can note the most intractable problem of the social scientist's life: bias in the experimenter. Much of the literature in our journals is concerned with exposing the alleged bias of either the experimenter or the data.

In part we create bias by approaching our experiments with a view to 'proving' something we are convinced about before we begin the experiment. It is not a question of finding, for instance, the level of real deprivation in a community, but of choosing a standard measure of wealth that either reflects massive or little deprivation, depending on which case we favour.

Sampling in theory and practice

We shall now return to the normal distribution and I open the discussion with a striking quotation from Galton, this time on *order in apparent chaos:*

> I know of scarcely anything so apt to impress the imagination as the wonderful form of cosmic order expressed by the 'Law of the Frequency of Error'. The law would have been personified by the Greeks and deified, if they had known of it. It reigns with serenity and in complete self-effacement amidst the wildest confusion. The huger the mob, and the greater the apparent anarchy the more perfect is its sway. It is the supreme law of Unreason. Whenever a large sample of chaotic elements are taken in hand and marshalled in the order of their magnitude, an unsuspected and most beautiful form of regularity proves to have been latent all along. The tops of the marshalled row form a flowing curve of invariable

> proportions; and each element, as it is sorted into place, finds, as it were, a pre-ordained niche, accurately adapted to fit it. If the measurement at any two specified Grades in the row are known, those that will be found at every other grade, except towards the extreme ends, can be predicted ... and with much precision (Quoted in Pearson, op. cit. p. 328).

This is much more than an enthusiastic eulogy of the normal curve, typical of a researcher's enthusiasm for his work, for it contains something germane to the topic of this section. Note the last sentence about being able to predict the placing of 'every other grade' in the arranged row on the basis of 'any two specified Grades'. Now this statement is pregnant with possibilities, for, assuming that the normal curve is all that Galton (and others) made it out to be, it was not necessary for a researcher to have the entire distribution of the data to hand in order to make predictions about how any given data conforming to the shape of the normal curve is distributed. The step from this implication to the modern techniques of statistics was a small one.

In other words, the researcher can make estimates of the distribution of a population on the basis of a sample from it. What an obvious and brilliant conclusion, and what a powerful one too! Let us see what mileage modern statistics has managed to make out of this simple but devastating implication.

The case for taking samples of a population as an alternative to enumerating all the data has already been discussed above, along with the case for applying the strictest of standards to the sampling procedure. Collecting the data is one thing, doing something with it is another. Having said that, we must also concede that hoping to do something with data without recourse to a mathematical theory of the operation is not likely to be productive.

In astronomy the normal curve of observational errors was known for decades before any applications of it to the social sciences were contemplated. Quetelet, you may remember, recalled the properties of the normal curve from his visit to astronomers in Paris when he began, some years later, to study the shapes of human beings and their behaviour.

Sociology needed more than the normal curve if it was to open up fertile areas of research into society. So did the budding sciences of genetics, its social offshoot, eugenics, and experimental biology.

Where astronomers had no reason to tread, the early statisticians had to. They needed a mathematics of random sampling in the context of a normal distribution that would enable them to establish the grounds on which they could assume that a sample was representative of the population as a whole. Out of this need grew modern statistics.

A random sample is a subset of a population. Our interest in the sample is in the degree to which we believe it is representative of that population. A random sample gives us an estimate of the properties of the population, and, naturally, we aim to get as accurate an estimate of those properties as we can. In doing so we are bound by the statistical 'rules of evidence', breach of which has a harmful effect on our reputations. Much of sampling theory in statistics is about those 'rules of evidence'.

Let us summarize an aspect of these 'rules'. Consider a sample of a population. It is a subset of the population and we want to know: what can we infer from the sample about the population?

The population will have a mean value and its standard deviation measures its dispersion about the mean. What of the random sample itself? It too will have a mean value, found by summing the values of the items in the sample and dividing by the number of items. It too will have a standard deviation. Is there any connection between the means and standard deviations of the population and the random sample? Yes there is, and the relationship is a most powerful assumption of statistics.

Note that there is no way of knowing the characteristics of a population for *certain* from a sample (even a large one). The only way to identify the characteristics of a population is to examine every single member of it (for example, conduct a complete census, or test to destruction the entire output of a plant and so on).

All samples by definition are smaller in number than the population itself, but statistics is not about certainties, it is about making inferences from limited data and measuring the degree of confidence that you can have in any specific instance about the likely properties of the population from which the sample is drawn. Suppose we took some samples randomly from a population (of what is not important, though how we select our samples is very important) and we compared the means of each sample. What would we find? That they were normally distributed!

If you think carefully about this, the logic of this remarkable relationship becomes obvious. There is no doubt that the sample means will vary slightly because we are taking items from a population that is itself distributed among a range of values. Because the distribution varies, so must the items taken from it and therefore samples bring their variations from the population mean with them.

Some items will pull their sample mean to the right of the population mean (more items in that sample are larger than the population mean than there are items smaller than it) and some items will pull their sample mean in the other direction. Imagine a large number of sample means, some bigger and some smaller than the population mean, and then find the average value of these sample means. What would you expect? Yes, that the mean of the sample means is *close* if not identical to the population or true mean.

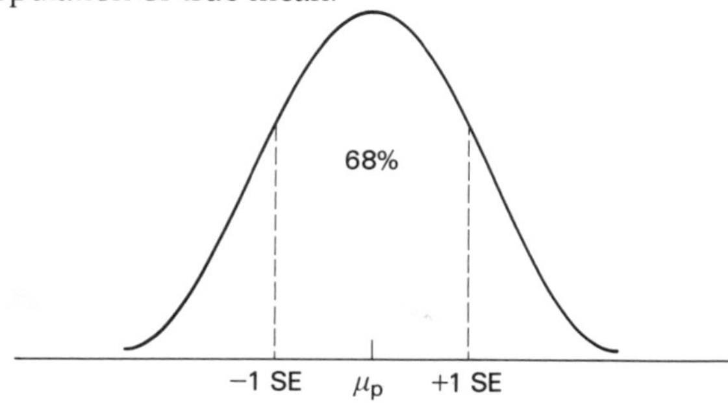

Figure 9.1 Distribution of sample means

The more samples we take the closer the mean of the sample means will be to the population mean, and, if the number of samples is large enough, the mean of the sample means will be the same as the population mean. In fact, the means of the samples will be normally distributed around the population mean because the chances of taking items for our samples close to the population mean is very much greater than the chance of taking items distant from the population mean. This is itself a property of the normal distribution (see figure 9.1).

Sixty-eight per cent of the items in the population will be found within one standard deviation of the population mean and there is roughly a two-thirds to one-third chance of picking an item from within this range rather than one from beyond it, in either direction. If items are biased in favour of being within a standard deviation of the population mean it follows that collecting items at random for our samples will produce sample means with the same bias: the mean of the sample means is a good approximation of the population mean.

What of the standard deviation of the sample means? This too follows logically from the points we have discussed above. The population distribution will be greater than the sample distribution, if only because the population distribution must include *all* the values of all the items that make it up, including those items towards the extremities (more than three standard deviations from the population mean), while, in contrast, the sample distribution is most likely to contain the great majority of its items selected (randomly of course) from within a single standard deviation of the population mean. In other words, the odds favour the standard deviation of the sample means being *smaller* than the standard deviation of the population as a whole.

SE's and SD's

To avoid confusion between the standard deviations of the population and the sample means, the standard deviation of the sample means is known as the *standard error of the sample mean,* or more simply as the standard error (SE).

The mathematics of the standard error are outside the scope of this book but some general points can be made, without proof. If our samples consisted of the entire population, then the sample mean would be identical to the population mean and there would be no variation between each sample mean (they would all be the same) and therefore the standard error of the sample means would be zero. Conversely, if we took successive samples of one item out of the population and found the sample mean for each one we would, in effect, be finding the population mean and its standard deviation, and therefore the standard error of the sample means would be exactly equal to the population standard deviation.

This gives us a range for the standard error of the mean, running from zero (sample of the entire population) up to the standard deviation of the population (samples of one). In practice, we are taking samples smaller than the population and greater than one (usually greater than 30 items at least) and therefore the standard error of the means will always lie between zero and the value of the population standard deviation. Just where it will lie will depend on the number of items in the sample – the greater the number of items the smaller the standard error.

Mathematicians have shown that the standard error of the sample means is actually equal to the standard deviation of the population divided by the square root of the number of items in the sample. While this result is not of great interest to you just yet, it is of great importance in statistics, for it relates the standard error of the sample means to the standard deviation of the population.

For practical purposes we can take the standard deviation of the sample to be such a good approximation of the standard deviation of the population as to use the former as if it was the latter. Thus, with the standard deviation of a single sample, we can use this to find the standard error of the sample means without having to actually find them directly.

The size of the standard error of the sample means reflects the dispersion of the population, but as long as the sample contains at least 30 items in it this is found in practice to be sufficiently representative of the population it is drawn from, irrespective of the variation within that population, and we can use the standard deviation in the sample as if it was the standard deviation in the population. (How we cope with samples of less than 30 items is discussed in chapter 11.)

Following from this point, the larger the number of items in a sample, the smaller the standard error of the sample means is likely to be, and the smaller therefore the range either side of the sample mean within which we expect the population mean to be contained (though we can never be certain).

However, there are diminishing returns to taking larger and larger samples from a population: beyond about ten per cent of a population the additional accuracy of the estimate of the population mean from our sample mean and standard error is unlikely to improve very much, hence, astonishingly, there is not much point taking samples in excess of ten per cent (and in practice we take much less than ten per cent) of a population.

This point is often taken up by people who see polling organizations making predictions about party preferences on the basis of what appear to be a very small proportion of the total electorate. They find it difficult to accept that only about 2,000 people in a properly conducted sample can be sufficient to infer party preferences of a voting population of 18 million.

It is not the proportion of the population that decides the accuracy of the results (though people who disagree with the poll results are inclined to believe it is) but the random *selection* of the sample and its absolute size (it must always exceed 30). If the sample is truly random, and properly structured, it will give results that are unlikely to change much if the sample is doubled, quadrapuled or even increased ten-fold.

So what?

We have leapt a long way from the early concerns of the statisticians of the 1900s and no doubt you are feeling slightly 'deviation drunk' at this moment. It is time therefore to bring together the assertions I have made and see what practical use they have.

For a start, everything rides on the properties of the normal distribution. Without the normal curve in your mind much of what makes sense in statistics will not! So let us go back to some of the normal curve's properties.

The area under the normal curve is neatly demarcated by varying probabilities that an item in a population will lie within a certain distance (in value) from the population mean. These probabilities express the degree of confidence we can have in estimating the relative value of an item in a population compared to other items and to the mean of all the items.

Sixty-eight per cent of the items in a population will fall within plus or minus one standard deviation of the mean value, 95 per cent within plus or minus two standard deviations of the mean value and over 99 per cent within plus or minus three standard deviations about the mean.

This gives us a method of being precise about our uncertainties in respect of particular populations. More, by relating samples to these populations (and others for which we can never hope to enumerate or test even if we wanted to, such as the population of all sea birds, or the output of light bulbs, or printed circuit boards and so on) we can make statements within precise limits of confidence about characteristics of populations on the basis of relatively small samples from them.

For example, on the basis of a sample mean we can estimate the likelihood that the population mean is within a specified range. We do this using the property that 68 per cent of the sample means will fall within the range of the population mean plus or minus the standard error of the means. Reversing this, we are saying in effect that there is a 0.68 probability that the true mean is within the range of the sample mean plus or minus the standard error (and a 0.32 probability that it is not).

For example, suppose that some students randomly selected out of a class of 100 have a mean examination mark of 62 with a standard deviation of 17. The standard error of the mean is found by dividing the standard deviation by the square root of the population which in this case gives us: $17/\sqrt{100} = 17/10 = 1.7$ marks. We can use this result to estimate the mean examination mark of the population of 100 students. Because of the properties of the normal curve, there is a 68 per cent probability that the population mean is within one standard error of the mean of the sample. This suggests that the population mean lies within the range of $62 + 1.7 = 63.7$ and $62 - 1.7 = 60.3$ (that is, between 60.3 and 63.7 marks).

A 68 per cent probability may not be good enough for our purposes and so we would have to increase the range which the population mean can fall. This is done by multiplying the standard error by 2 for an estimate within the range of two standard errors and by 3 for a range of three standard errors. The arithmetic is easy: $2 \times 1.7 = 3.4$ and so range for 95 per cent confidence is therefore $62 + 3.4 = 65.4$, and $62 - 3.4 = 58.6$ (or 58.6 and 65.4); for a 99.7 per cent confidence we would get $3 \times 1.7 = 5.1$ and taking this either side of the sample mean of 62 marks we get: $62 + 5.1 = 67.1$ and $62 - 5.1 = 56.9$ (or 56.9 and 67.1).

In making this calculation we ought to remember that it is possible that the population mean lies outside this range: at the 68 per cent level there is a 16 per cent chance that the population mean is greater than 63.7 marks and a 16 per cent chance that it is less than 60.3 marks. By extending the range to increase our confidence that the population mean lies within two standard errors of the sample mean mark we diminish the chances that it lies outside the range. At two standard errors the chances of it being outside the range of 58.6 and 65.4 marks is 5 per cent. At three times the standard error the chances are reduced to 0.3 per cent only, but remember this could still occur. All that we have done using this method is tell you just how confident you *can* be about where the population mean is in relation to the sample mean; how confident you *want* to be is entirely up to you!

10

Testing, Testing

Karl Pearson

Among the giants of the natural sciences and mathematics the publication work rate is often much greater than in the social sciences. In the case, however, of Professor Karl Pearson (1857–1936), he published over 600 articles in his working career, and in doing so left his mark in modern statistics.

Pearson, according to contemporary accounts, was not the easiest man in the world to work with, or for. He had a predilection for a strange mixture of passionate beliefs and scientific rigour. Consider his intellectual background: he had the classical education of a Victorian scholar – Kings College, Cambridge (where he became third wrangler in 1879), followed by a year at Heidelberg and Berlin universities, and then called to the bar in 1881. Where he began to diverge from the normal track for young scholars was in his part-time lecturing on Karl Marx's theories to those thirsty for knowledge on a Sunday in London.

It is suggested by Haldane (1957) that he changed the spelling of his first name from a C to a K in tribute to Karl Marx, and we know he believed that socialism would emerge as the state, and what we could call today the technocracy, gradually took over the functions of the bourgeois entrepreneur. In him we have a sign of his times: a Fabian belief in the inevitability of gradualism.

In these early years, Pearson published a couple of books anonymously, which echoed his earlier anti-religious convictions. At Kings he had refused to attend divinity lectures, and his first books continued his undergraduate stances on official religion with an attack on Christian orthodoxy.

Like other intellectuals of the nineteenth century (for example, John Stuart Mill) Pearson was a feminist and he joined the 'Men's and Women's Club', where, among other things, matters of sexual freedom were discussed (though by all accounts not practised, it was after all the 1880s not the 1980s; in fact he married one of the club's members).

In sum, Karl Pearson was a person of strong early beliefs and was oddly left of centre in political, social and religious matters. Apparently, his radical views did not lead to his exclusion from the teaching institutions of the day (as happened in Sweden to the brilliant economist, Knut Wicksell) and he was appointed to a chair in applied mathematics and mechanics at University College, London at the age of 27.

From then on, Pearson devoted himself to academic work, and while his contributions to statistical methods have lasted into modern textbooks, his own personality and ideas about the world have long since had that peculiarly English treatment: where something cannot be forgiven it is simply forgotten.

Of course, Pearson's connections with the eugenics movement has not been entirely forgotten, nor could it be forgotten as he held the Galton Chair in Eugenics from 1911 to 1933, and much of his work was concerned with the study of hereditary issues.

Pearson was much more than a crude advocate of 'racial' improvement and his views ought not to be confused with racism. He and his colleagues wanted to improve the human race biologically by reducing the proportion of births that transmitted heriditary defects. To give credence to this high moral goal they sought a science that could bring it about, and for this they needed data.

Even the word racial had a more innocent connotation in his day compared with ours, being used as a synonym for 'species'. The interest of social reformers in the issues of inter-generational inheritance were the inevitable result of the Darwinian revolution catching the imaginations of those who sought nought but progress for the human species. To condemn Pearson is to condemn many of the intellectual lights of the early Labour movement.

Pearson's anti-religious views, one may surmise, inclined him towards a theory of evolution that placed it within the natural rather than the divine order of things. But much remained to be done. Darwin's theory of natural selection was an untested hypothesis in search of a means of confirmation or rejection, on other than purely ideological grounds, and the Darwinian research agenda fascinated the best and the brightest minds of the age.

Pearson was among those young men who threw themselves into the study of inheritance with the enthusiasm common to youth on the crest of a revolution. Similar behaviour was exhibited by young economists when Keynes published his *General Theory* in 1936; and Wordsworth caught the mood well with his lines about it being 'bliss to be alive and to be young was very heaven' in reference to the hopes inspired by the early years of the French Revolution in 1789.

All a question of arithmetic

Those interested in the study of inheritance literally had to invent the means to test Darwin's conjectures, for no readily available science existed that could cope with their requirements. Statistics had not yet achieved anything like the level found today in an elementary textbook, and the necessity of finding a way of measuring things that had not been measured or compared before was itself a powerful motive for the discovery of statistical techniques.

Scientists had many questions, most without answers or the means of acquiring them, and in this way the questions led to the new means which led to new answers. The overall strategic question was clear: was it possible to observe natural selection by examining the apparent characteristics of plants and insects in different environments? If it was, then the power of empirical evidence could be joined to the sense of Darwin's conjectures.

To observe natural selection required the careful measurement and recording of large numbers of a species, and, having collected the data, there was the problem of how to assess the meaning of small differences in physiological characteristics.

In effect, the scientist had to decide how much of the

difference in a common characteristic of a species was due to variation within the same variety of species, and, by implication, at what point a difference in a characteristic indicated the existence of different varieties of the same species?

The answers to these types of questions were not easy to find, largely because of the separation of mathematics and science. The answers were not in themselves difficult; just that nobody had asked them before. As the questions themselves had not been posed before, the answers either had to come from current mathematical knowledge or by the creation of new statistical methods.

The posing of questions, such as those that inspired Galton and his colleagues, including Pearson, led them to mathematical notions about the data and the invention of the modern statistical approach.

Galton and Pearson founded the journal *Biometrika* in 1901 and devoted it to the study of inheritance and the environment. By its nature, *Biometrika* became the journal of statistical method and in its pages much of the really important developments in statistics in the early years is to be found.

A single paragraph from its first editorial shows, effectively, how its intended role led to its lasting contribution to fields much wider than the interests of its first subscribers and contributors:

> The starting point of Darwin's theory of evolution is precisely the existence of those differences between individual members of a race or species which morphologists for the most part rightly neglect. The first condition necessary, in order that any process of Natural Selection may begin among a race, or species, is the existence of differences among its members; and the first step in an enquiry into the possible effect of a selective process upon any character in the race must be an estimate of the frequency with which individuals, exhibiting any degree of abnormality with respect to that character, occur. The unit, with which such an enquiry must deal, is not an individual but a race, or a statistically representative sample of a race; and the result must take the form of a numerical statement, showing the relative frequency with

> which the various kinds of individuals composing the race occur (*Biometrika*, vol. 1, p. 1).

Note the use of recognizably statistical terminology in that paragraph: *estimate; frequency; degree of abnormality; statistically representative sample; numerical statement;* and *relative frequency*. The biometricians had come a long way since the efforts of John Graunt. Once the biometricians discovered the appropriate methods for their field of interest, they opened the way to applications of the same methods in other fields, and they also provided a two way movement between themselves, as empirical scientists trying to answer questions peculiar to their own interests, and to others who were mathematicians. The latter were often asked to apply themselves to the kinds of problems that interested the new school of statistics.

The work of the biometricians was well under way before the first edition of *Biometrika*. Professor Walter Weldon, holder of the Jodrell Chair in Zoology at University College, had been busy measuring the *Decapod Crustacea* (a shrimp to the rest of us!) in 1890.

He discovered that the distribution of variations in the shrimp were similar to a normal distribution. This supported work by Galton on the sweet pea and Merrifield on the size of moths. Once the scent was up, the pack took off with boundless energy.

Weldon quickly took to this work – he even took time out as a Professor to learn some mathematics, which some readers may wish to emulate – and he provided a rich harvest of results for the statistical researches of the likes of Pearson, who had begun the first ever university lecture series on statistics as a subject in its own right in 1894 for students at University College, London.

A final quote from Weldon will set the tone for this chapter. In it Weldon discusses the tasks facing the biometricians if they are to make useful statements about their work that will have wider interest:

> It cannot be too strongly urged that the problem of animal evolution is essentially a statistical problem: that before we can properly estimate the changes at present going on in a race or species we must know accurately (a) the percentage

> animals which exhibit a given amount of abnormality with regard to a particular character; (b) the degree of abnormality of other organs which accompanies a given abnormality of one; (c) the difference between the death rate per cent in animals of different degrees of abnormality with respect to any organ; (d) the abnormality of the offspring in terms of the abnormality of parents and vice versa. These are all questions of arithmetic; and when we know the numerical answers to these questions for a number of species we shall know the deviation and the rate of change in these species at the present day – a knowledge which is the only legitimate basis for speculations as to their past history, and future fate (quoted in K. Pearson, 1906).

It was the search for these solutions that were 'all questions of arithmetic' that was the decisive contribution of Galton, Weldon, Pearson, Gossett, Fisher and others.

The theories of heredity that motivated some of the leading individuals in the modern school of statistics are of less consequence to the search for a scientific statistical method than the scientific methodology they were led to in their quest for the truth. As Haldane put it (op. cit. p. 303), the theory of heredity may have been 'incorrect in some fundamental respects' but 'so was Columbus' theory of geography. He set out for China, and discovered America'.

'Student'

William Gosset (1876–1937) enters our story for an impeccable reason: his major contribution to statistics is still taught in statistics courses (have a look for his 't' distribution table in your textbook). He made many other contributions to statistical methods (always under the self-effacing pseudonym 'Student', indicating a modesty that is rare among those who have stalked successfully along the frontiers of knowledge).

As if statistics does not have enough to contend with through its association with gambling, those who disdain the imbibing of booze will now discover one of its most talented contributors producing the world famous Guinness stout. (Without conceding to either of these prejudices, all I can say is, thank heavens that the likes of Florence Nightingale also contributed to statistics!)

Gosset graduated from New College, Oxford, in chemistry and mathematics and went on from there to work as a brewer for the Guinness Company in Dublin in 1899. His work and early mathematical training led him to concern himself professionally with practical problems for his company, and, inevitably, this led him to Karl Pearson and *Biometrika*.

One of Gosset's principal interests was how to make sensible statements about the properties of small samples. By way of contrast, the biometricians around Pearson concerned themselves with large samples. If ten thousand shrimps were available for measurement, out of a population of millions, why confine yourself to studying a couple of dozen? For Guinness the problem was entirely different. Field experiments with varieties of barley are costly enough in time and resources without trying to undertake hundreds of them.

For the brewery's purposes, it was (and had to be) enough to test the barley varieties in a few field plots – perhaps as little as a handful – and to make decisions on the basis of the results, for

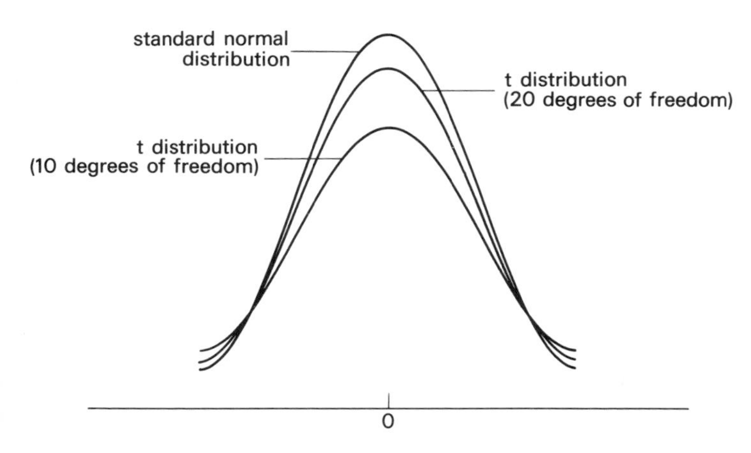

Figure 10.1 *Standard normal and t distributions*

the brewery existed to sell consistent quality stout, preferably at a profit, not to fund pure research programmes into evolution.

Restricting the sample size creates its own problems. For a start, while it is satisfactory to assume that the population is approximately normally distributed about its mean, it is certainly unsatisfactory to assume that the distribution of *small* samples from a population is normally distributed. By small samples we think of numbers less than 30.

If the small sample distribution is not normally distributed, how is it distributed? That depends on the size of the small sample. What Gosset suggested was that the distribution could be assumed to be symmetrical about its mean but that its shape would diverge from the normal distribution and that this divergence had to be taken into account when trying to make statements about the population mean and standard deviation.

The shape of the small distribution is related to the normal distribution by the size of the sample; as the sample increases the small sample distribution approaches the shape of the normal curve:

> if the sample is very small, the 'tails' of its distribution take longer to approach the horizontal and its 'head' is below the 'head' of the normal distribution; as the number in the sample increases, the tails move 'inwards' towards the tails of the normal distribution and the 'head' of the curve moves upwards towards the 'head' of the normal distribution (see figure 10.1).

What Gosset did was to specify the nature of the shapes of the small sample distributions and to represent them in the form of a table of their standard errors for each sample size.

This will be clearer if you recall what is involved in estimating the population mean from large samples. If the sample means are normally distributed, then we know that the population mean lies between the estimated mean plus or minus 1.96 times the standard error of the estimated mean. This is for a 95 per cent confidence level; it is within plus or minus 2.58 times the standard error at the one per cent confidence level (chapter 9).

If, however, the sample is small and its distribution diverges from the normal because of this, we are less sure of the interval within which we can reasonably expect to find the population mean. Therefore, for the same level of confidence we must *widen* our estimates of the interval within which the population mean might be expected to fall. This follows from the 'flatter' shape of the small sample distribution (see figure 10.1). In other words, we are trying to cover the risk that the small samples are less representative of the population by widening the range within which we believe the population mean will fall.

Instead of multiplying the standard error by 1.96, at a 95 per cent confidence level, we must multiply it by a larger number. How large the number depends upon the size of the sample–the smaller the sample, the larger the number.

For instance, for a sample of about 30 we must multiply the standard error by 2.04, not 1.96; for a sample of about 20 the multiplier is 2.09; for one of about 10 it is 2.23; and for one of about 5 it is 2.57. The multipliers are found in a 'Student's t' table and they are given for various confidence levels.

Degrees of freedom

At this point we must introduce ourselves to a curious (at least to beginners) notion, known as the 'degrees of freedom'. Nothing causes more confusion to beginners in statistics than the idea of the 'degrees of freedom'.

Suppose we are drawing very small samples from a population and calculating their means: the smaller the samples the wider the variation there will be in their means; the larger the samples the smaller the variation in their means.

This should appeal intuitively to you: if I take random samples of two students from the university population and compare their mean heights with the mean heights of the same number of samples of 50 students, which set of samples would you expect to have the largest variation?

A single student from the extremes of the height distribution in my samples of two would throw up a large variation in the mean height of the sample, while in the sample of 50 students, a single student from the extremes is more likely to be affected by those nearer to the norm height of the population among the other 49 students in the sample. This will tend to reduce the variation in the larger samples.

Suppose we found that in a small sample of four items we had a mean of 25 (what we are measuring is irrelevant here). The total in this case must be 4 times 25 equals 100. Now no matter what the actual values of the individual items are we are *free* to give them any values we like as long as they sum to 100 – because only if they sum to 100 can they be divided by 4 to get a mean of 25.

If we give three of the four items the values of 25, 29, and 22 and sum them we get a total of 76. In giving the items these values we automatically *restrict* our choice of the value of the fourth item – our freedom is reduced. The value of the fourth item under these conditions is simply 100 – 76 = 24; it can have no other value, for only 76 + 24 = 100. In other words, we are *free* in three cases out of the four but are restricted in the fourth.

We measure our degree of freedom by taking the case we are restricted in from the total items available, and this clearly is 4 – 1 = 3. Thus, in determining the mean of a sample, we call this three degrees of freedom, usually identified as N–1 degrees of freedom.

In the case of a large sample the N–1 degrees of freedom are still operating, but with a large number (over 30) the square root of N is not much different from the square root of N–1. If the sample size runs into hundreds or thousands the difference is so small as to be of no consequence.

Exercise 10.1

Test this with a pocket calculator by finding the square roots of 8 and 8–1 = 7 and of 8,000 and 8,000–1 = 7,999 (see Appendix).

In calculating the standard error (see chapter 9) for a small sample we must divide the standard deviation of the sample means not by the number in the sample (N) but by N–1 to indicate the loss of a degree of freedom, that is, we cannot have any values we like for *all* the items in the sample, but we can for one less than the sample number. The effect of dividing by N–1 instead of N is to make smaller the number we are using to divide the sample standard deviation, and this must make our result slightly bigger, and thus widen out the range within which the population mean is likely to be found.

The fact that we allow a slightly wider range of values for the population mean, when we must rely on a small sample, reflects our concern that we are not misled by the small sample mean into placing too much confidence in the possibility that the population mean is as close to the sample mean as it would be if we had used a much larger sample.

'Tis only an alternative hypothesis

Statistics is fortunate in that during its most formative period (roughly from the turn of this century to the Second World War), the lives of a number of its outstanding exponents were enmeshed to a greater or lesser extent in roughly overlapping periods. Among the positive features of this fortuitous accident of history is the obvious one that their creative work in statistical theory had to run the gauntlet of scrutiny by their rivals and contemporaries.

The negative feature was distressing for those involved – scientific rivalry too often slipped into petty disrespect for, and consequent disregard of, each other's work – and a lot of unnecessary stress was generated in petty sulks and quarrels.

This sort of approach to science is not always conducive to its speedy progress, particularly if proponents of a point of view stubbornly refuse to adapt or correct their methods in the light of other peoples' contributions.

Karl Pearson's son, E. S. Pearson, recounts (1966) the story of how the geneticist, William Bateson, at the 1904 meeting of the British Association for the Advancement of Science, dramatically threw onto table all the then issues of his father's journal, *Biometrika,* with the remark 'I hold that the whole of that is worthless'. Other 'atrocity' stories abound in the discipline, not the least of them emanating from the career of Ronald A. Fisher (1890–1962) and his disputes with Karl Pearson.

Fisher was a prodigous mathematician and made early contributions to *Biometrika* (notably in 1912 and 1915). He established mathematical proofs of the important statistical theorem that was developed by 'Student'. Pearson's editorship of *Biometrika* effectively set him up as the arbiter of scientific excellence and it must be said that his treatment of the young Fisher did not make any concessions to scientific disinterest. In short, he effectively blocked some important contributions from Fisher, and worse, published some criticisms of Fisher's work that were grossly unfair to the younger man's reputation.

The details (see Joan Fisher Box, 1978) need not detain us, as we want to look at some of the consequential events in statistical methodology that separated the profession for many years (and which are not entirely resolved yet).

However, Fisher's own comments, looking back from the peak of his professional reputation in 1947, deserve to be studied by readers who are shocked by, what is crudely called, 'academic bitchiness' when they witness it. These comments might be all the more valuable if, in the future, you ever feel its cold blast across your own discoveries:

> A scientific career is peculiar in some ways. Its *raison d'être* is the increase of natural knowledge. Occasionally, therefore, an increase in natural knowledge occurs. But this is tactless and feelings are hurt. For in some small degree it is inevitable that views previously expounded are shown to be either obsolete or false. Most people, I think, can recognise this and take it in good part if what they have been teaching for ten years or so comes to need a little revision; but some undoubtedly take it hard, as a blow to their *amour propre,* or even as an invasion of the territory they had come to think of as exclusively their own, and they must react with the same ferocity as we can see in the robins and chaffinches these spring days when they resent an intrusion into their little territories. I do not think anything can be done about it. It is inherent in the nature of our profession; but a young scientist may be warned and advised that when he has a jewel to offer for the enrichment of mankind some certainly will wish to turn and rend him (quoted in J. Fisher Box, 1978, p. 131).

Characteristically, Fisher had his 'revenge'; having refused, while in great need of a salaried post, an appointment under Pearson at his Galton Laboratory in 1919 (suspecting the 'independence' of Pearson's leadership), he succeeded to Pearson's Chair in 1933.

Out of the debates on Fisher's scientific work, two schools of thought can be said to have emerged: his own, based on rigorous mathematics and seeking to illuminate the potential inferences in a set of data (the 'maximum likelihood' and 'fiducial probability' theorems), and that of E. G. Pearson, Jerzy Neyman and Abraham Wald (1902–1950) (hypothesis tests, probability frequencies and decision theorems).

Inference versus decision

These differing approaches were essentially different interpretations of the significance of data under uncertainty. Perhaps the essence of the debate can be got across with a simple statement of the alternative consequences of examining a given set of data.

A Fisherian statistician would seek to estimate the probable reliability of an hypothesis, that is, to what degree of probability is the hypothesis likely to be true given the actual data that is available? This probability value is not a statement that the hypothesis is confirmed, or refuted. It is solely an estimate of how likely it is as an hypothesis.

If the hypothesis has a high probability that it is consistent with the data, that is all that the statistician reports. If there is a low probability, the Fisherian statistician would record either that the hypothesis was false, *or* that an event with a low probability had occurred, and he would make no *decision* as to which was the case.

In illustration, consider the case where a student requests more time to complete an assignment, and research suggests that only 10 per cent of students who require an extension tell the truth about the reasons for wanting one. The Fisherian statistician would report that the student is either a liar or a saint, and that an event with a 0.10 probability (that he is a saint) had occurred.

The statistician's report would be confined to his estimate of the likelihood of the moral status of the student. It would not extend or imply anything else that is allegedly consequential of the estimate. For instance, it does not follow from the estimate that there is a 0.9 probability that the students who hand in their assignments on time are saints and not liars – they may very well be liars but having completed their assignments on time (for whatever reasons) they have not needed to exercise their talents on this occasion.

The alternative to the Fisherian concept of statistics as an inferential exercise, is that associated with the names of Pearson-Neyman-Wald. Their approach is much more decisive than Fisher's: they go in where Fisherians fear to tread. They believe it is legitimate to use the statistical treatment of data to make a decision about the validity or otherwise of an hypothesis, and they are prepared to take a relatively small risk that they will make the wrong decision, rather than not make a decision at all.

Fisher saw statistics as being about inference from uncertain data, while most modern statisticians follow the decision school, opting for making an optimal judgement under conditions of uncertainty. Fortunately, both approaches are taught today, though, unfortunately, the philosophical debate has been squeezed to footnote status, or is picked up indirectly by means of cautions to users of the popular decision rules discussed below.

Decision statistics prescribes strict rules for making decisions and it is to these that we now turn.

11

To Err is Human

Scientific inquiry can be seen as a progression of guesses, running from wild through to inspired, that get ever more closer to the 'truth' as knowledge accumulates (sometimes overthrowing current truths in favour of new insights and new 'truths'). Just as artillery gunners find their targets by successive approximation – firing ranging shots, some short, some beyond, and some with left or right drift, until they can fire for effect on the most accurate bearing – so we can think of scientists creating conjectures about the world and testing them against data.

Scientific guesses are what we describe as hypotheses (unscientific guesses are likely to include a high dosage of belief in explanations that are superstitious). With experience and greater understanding of their subjects, scientists get better at formulating a likely hypothesis, much as an experienced gunner learns to quickly find the accurate range for his gun in whatever conditions he is called upon to fire it.

Much of the work of serious social science is concerned with stating potential hypotheses in ways that enable testable predictions to be made. By setting them out this way it is possible to use empirical evidence to examine their probable falseness. Hypotheses that are falsified by evidence are, or ought to be (though not always in practice it is sad to acknowledge) rejected in favour of those that survive the most rigorous of tests.

It is the test of the soundness of a theory that it passes all attempts to discredit it with the evidence of the real world. This viewpoint – that hypotheses should be tested against the real world – is not unanimously shared by all who prescribe doctrines for how society should be organized (whether they plan to leave the world as it is, or to change it is not important; both radical and conservative doctrines can depend on the falacies of untestable programmes).

The reason why some people ('scientists' included!) prefer to work with or believe in theories that depend upon untestable propositions is simply that an assertion that admits to no tests of experience has a unique advantage over those that do: they can never be refuted.

What cannot be refuted is immune to rational challenge and it becomes an article of faith, whether it is concerned with the simultaneous perfectability of mankind, the permanent infallibility of a leader, the objective and inevitable malevolence of a ruling class or ethnic minority group, or a belief that the good intentions of reformers are sufficient to guarantee the validity of their prescriptions.

Assuming that a scientific approach to hypothesis formulation and testing is accepted as the only scientific approach to a discipline, there still remains the problem of establishing a set of rules for conducting the tests. These must be at least as rigorous as the rules for conducting the experiment or enquiry to collect the data (discussed in chapter 10).

Types of Errors

Jerzy Neyman and Egon Pearson formed a fruitful scientific collaboration between 1926 and 1934, and though often separated by their countries of residence (Neyman in Poland and Pearson in Britain), they produced some remarkably important work, particularly on the subject of testing hypotheses (they titled the article that we are interested in: 'On the problem of the most efficient tests of statistical hypotheses', which, as befitted a pathbreaking step forwards for statistical science, was published in the prestigious *Philosophical Transactions of the Royal Society* in 1933).

The gist of their approach was to set out the conditions under which it would be 'safe' to accept or reject a hypothesis given the data that had been collected and processsed. In setting out formal rules they were offering the profession a standardized approach to the problem that beset all hypotheses when they were confronted with the data: does the data support or refute the hypothesis?

Unless rules could be agreed – and to be agreed they have to pass the critical scrutiny of the profession – there was the risk that all kinds of arbitrary standards would be set, and not all of them scrupulous of the evidence where it might suit an 'enthusiastic' or over-committed researcher to protect a position with 'soft' rules.

Over the years the Neyman–Pearson theory of hypothesis testing has been developed into a formal theory that forms an important methodological part of statistics. I refer to the method of testing an hypothesis by means of attempting to disprove the *null hypothesis*.

In estimating procedures we attempt to find the probablilty that the true value of the population parameters (mean, standard deviation) are within a given range of the sample parameters. In hypothesis testing we want to make one of two decisions: either we accept the hypothesis or we reject it. This is a different procedure with different implications. At the most obvious level it is decisive: we either do one thing or another. We do not leave it to the reader, or our colleagues, to choose; we do that for them.

If others are sceptical of our decision, they have the option of checking our calculations (and there is always a risk of error in the arithmetic – hopefully for the confidence of our colleagues in our reliability, this risk ought to be small, but never get too cocky about that!), or they can repeat the collection of data and apply the same tests to it to see if we have accepted a sample that is not representative of the population.

It is important to distinguish here between the fact that though our decision is decisive (we have chosen to accept or reject an hypothesis), the statistical test itself is *not* decisive. The test is an accepted convention for making a decision, it is not a guarantee that the decision itself is a correct one!

The formal statement of the problem is to decide on a hypothesis and to test it against the evidence. This in fact involves two competing hypotheses: one hypothesis that something asserted is true, and another that it is not true.

Do miners beat their wives?

Consider an example: suppose I was to hypothesize that there is a difference between the amount of domestic violence in households where the male parent is a coal-miner compared to that found in the rest of the population, in other words, coal miners beat their wives more frequently than other male occupation groups. (I hasten to add, for the benefit of my many friends in the coal industry, that I have absolutely no reason to believe this assertion to be true.)

In statistical terms I am asserting that there is a significant difference in domestic behaviour between the two occupation groups. The hypothesis that there is no significant difference between the two groups is known as the *null hypothesis.* This asserts that the amount of domestic violence in coal miners' households is no different from that found in any other occupation group, and that any small differences found in samples from both groups are due to chance influences.

The null hypothesis states that coal miners come from the same population as other occupations as far as the propensity to domestic violence is concerned. It follows that the *alternative* hypothesis asserts that they do not come from the same statistical population, that they are in fact two separate populations as far as the propensity to domestic violence is concerned, that is, coal-miners do beat their wives more frequently.

The issue could be important if we are going to decide between concentrating scarce counselling resources on a particular occupation group (believed to have a higher propensity to domestic violence than the general norm), and spreading the scarce resources across all occupation groups.

To resolve this issue using the decision rules, we have to attempt to disprove the null hypothesis, given that several confirmations of an hypothesis can never have the same weight as a single instance of its rejection. If the null hypothesis survives our attempts to discredit it, we have a good reason to take a risk of being wrong and by accepting this risk reject the alternative hypothesis; if it does not survive our tests, we have good reason to take a risk of being wrong and reject it in favour of the alternative hypothesis.

The risk of being wrong in each situation is very real. The usual analogy is that of deciding whether an accused person is innocent or guilty of a particular crime. In legal systems where juries are used, the prosecution has to prove 'beyond reasonable doubt' that the accused is guilty if the jury is to be persuaded to return a guilty verdict.

If the prosecution fails to do so to the satisfaction of the jury (not themselves!) it may be that a guilty person will be released back into society. This is a risk that cannot be avoided, if there is to be a very small risk of an innocent person being found guilty and sentenced to some wholly undeserved punishment.

There are two risks involved: that an innocent person will be found guilty, or a guilty person will be found innocent. Where their lives are at stake, there can be appalling consequences if we decide that innocent persons are guilty and execute them as a result. Likewise, there can also be appalling consequences if we decide that guilty persons are in fact innocent, and upon release they violently molest other victims.

If the null hypothesis is rejected when in fact it is true we have what is called a *Type I error.* In our imaginary example, we risk a Type I error if we reject the null hypothesis when it is in fact true that coal-miners in respect of domestic violence are no different from other groups.

If we claim that miners are different when in fact they are not, we may make a wrong decision to allocate our scarce counselling resources to miners' households (and also cause offence unjustly to the mining communities). We might also get wholly undeserved professional credit – promotion; prizes; research grants; increased sales of our books; publication of our monographs and articles; and so on – for 'discovering' something that is in fact false.

The profession tends to be conservative in dispensing credit and has a bias in favour of minimizing Type I errors. To minimize the risk of a Type I error we can increase the severity of the test, and one way to do this is to require the differences in the sample parameters to be *statistically* significant for 99 cases out of a 100, or even, 999 out of 1000 (see table 11.1).

Before discussing the significance of statistical significance (itself a controversial topic), we should note that the decision rule does not eliminate the risk of error; it specifies what the risk is and permits a decision to be made if the risk is demonstrated to be 'small' enough. The more stringent the test, the less likely we will make a Type I error, but in avoiding a Type I error we increase the risk that we will make a *Type II error.*

		Null hypothesis: True	Null hypothesis: False
Decision	Reject	*Type I error*	*Correct*
	Accept	*Correct*	*Type II error*

Table 11.1 Type I/II error table

Perhaps miners do beat their wives more often than non-miners (the null hypothesis is untrue). If we incorrectly accept the null hypothesis when it is untrue, we are making a Type II error, and in consequence we may direct scarce counselling resources to all households, thus spreading the effect more thinly (that is, reducing domestic violence less efficiently), when in fact we should concentrate our scarce resources on miner's households to decrease the incidence of domestic violence experienced by a particularly disadvantaged group of wives.

How significant is statistical significance?

Considerable confusion obtains in respect of the use of the word 'significant' in statistics. Without care, it could be taken that a hypothesis declared to be 'statistically significant' somehow attains the status of being important, or relevant in some way, or of special interest to decision-makers, or of note to researchers. It might have these attributes and deserve them; on the other hand it might not. This can be more easily appreciated if we outline the steps that are considered to be normal if a test for statistical significance is to be credible.

The researcher starts by formulating a hypothesis. The particular hypothesis may arise from the researcher's own programme of work, or a 'hunch' of some kind, or from published work that provides an opportunity for investigating comparabilities (for example, the comparative behaviour of delinquents in the UK and the USA), or for criticizing its findings (for example, the money supply as the cause of inflation).

Researchers are required to specify what they would regard as sufficient evidence for or against their hypothesis, implying by doing so that they would tentatively accept or decisively reject their hypothesis on the basis of the evidence only. It is expected of them that they specify the tests and the standards they would expect to reach before they undertake the tests. They are not expected to calculate first and specify later, for this could lead to *ad hoc* adjustments of their hypothesis in the light of the data. Just how much notice is taken of this requirement in practice is difficult to judge. From anecdotal experience I would suggest that it is observed less often than the profession would like.

The researcher's next step is to undertake data collection and to process the numerical inputs through commonly acceptable statistical procedures and subject them to clearly defined criteria. The data that is used in a piece of work is generally summarized in the researcher's published reports, and where necessary, readers are also invited to consult the raw data. If the data is from a published source, identification of that source is sufficient to enable others to check on the work, if they are inclined to do so. The possibility that somebody might re-work the data is always a constraint on sloppy or misleading use of data, there being nothing more embarrassing than having one's conclusions challenged on the basis of one's own sources!

On the basis of the processed data, and the pre-determined decision criteria for tentatively accepting or rejecting the hypothesis, the researcher can make a decision about the probable validity of the hypothesis (that is, whether to reject it as being inconsistent with the evidence and the standards set for it).

To reduce the possibility of a Type I error (it can never be eliminated) the researcher is required to set stringent tests of the data; the more stringent the tests the less chance there is of a Type I error. At this point the the level of significance is introduced.

A significance level specifies what the chances are of an event or parameter value occurring. A 5 per cent level of significance says that there is one chance in twenty that an observed value can occur and still be considered to come from a sample of the population in question. A 1 per cent level of significance tells us that there is only one chance in a hundred of this being the case; a 0.001 per level of significance says it is only one chance in a thousand of being true.

The smaller we require the level of significance to reach, the smaller the chance of a Type I error, because, if the manipulation of the data shows that the null hypothesis has less than one chance in a thousand of being true, it follows that the parameter value has a very small chance of coming from the population we are considering. Therefore, if we reject the null hypothesis under these conditions there is a very small (in the sense of being broadly acceptable) risk that we will commit a Type I error and reject something that is in fact true.

If we reduce the risk of a Type I error we increase the risk of a Type II error (rejecting the alternative hypothesis when in fact it is true). It may well be that the data manipulation shows that there is a relatively high chance that the sample is from the population we hypothesize it comes from, but this could be because our sample selection by chance includes a high proportion of items from the extremes of another population. That one chance in a thousand can come up!

The only way to reduce the chances of either Type I errors and Type II errors (they can never be eliminated) is to increase the size of the sample, taking care, of course, to ensure that the selection is random and not biased and that each item is selected independently of every other item.

It is in this context that the statistical significance should be understood. It is a technical device to decide whether a specific hypothesis meets a given standard. In and of itself it *proves* absolutely nothing (the thousand in one chance can occur) and the conclusions may have little relevance in fact.

The demonstration that a particular hypothesis has passed the stringent tests set for it by the researcher is a 'code word', if you like, to other researchers that there may well be something in the hypothesis, and that this should be tentatively noted until additional tests show that similar samples are as robust.

Even then, no amount of tests can prove any hypothesis to be true (Hume's dilemma), though a repeated failure of a prediction to materialize in the form set for it in the hypothesis should be sufficient to cast grave doubts on any hypothesis. Some would argue that a failure of a prediction requires an hypothesis to be severely amended, if not rejected outright.

One and two tail tests

We must look a little closer at the null/alternative hypothesis dichotomy: if we reject one we imply an acceptance of the other. This is fairly straightforward for many cases where we are testing the null hypothesis, with the explicit intention of trying to reject it – for it is assumed that our conjectured hypothesis is spelt out in the alternative hypothesis.

It is less straightforward in those other cases where, while the null hypothesis is clear, it is not exactly clear what should be the alternative hypothesis. This boils down to deciding not just whether a sample is the same or different from a population but also in which direction it is different. For example, a quality control system that samples items for test against certain standards might be required to reject an entire batch if the rejection rate is significantly different from 5 per cent.

The null hypothesis would be that the rejection rate in a batch would equal 5 per cent and therefore that the batch could pass if this rate obtained; but what would be the alternative hypothesis? Would it be to reject a batch if the rejection rate is less than 5 per cent, or more than 5 per cent? After all the sample could be significantly different from the standard in either direction.

In practice we would expect the rule to be to reject a batch if the rejection rate *exceeds* 5 per cent, that is, that there is a signficant difference between the sample and the norm for that population if the rejection rate was significantly higher than 5 per cent. It could be significantly lower, but we would normally ignore such a fortuitous event.

Suppose I was to hypothesize that students who read this book will score significantly differently in their statistics examinations from those who do not. The null hypothesis is obvious: there will be no significant difference in the examinations scores for both groups of students (that is, reading the book does not influence examination performance).

There are, however, two possible alternative hypotheses to the null hypothesis: either students who read the book will do significantly better than non-readers, or that non-readers will do significantly better than readers. This corresponds to the tails of a normal distribution. If readers do better then they are members of a different population to the non-readers and the mean score of readers is to the right of the mean score of the non-readers; if readers do worse, I am suggesting that their mean score is lower than the mean score of the non-readers.

In this case I reject the null hypothesis if the alternative hypothesis shows that its mean is outside certain limits set for the null hypothesis, that is, if it falls in either the right or the left tail of the normal distribution (see figure 11.1).

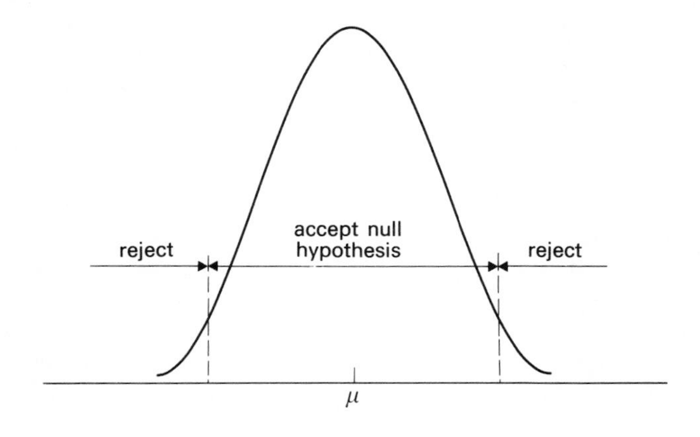

Figure 11.1 Decision rule for a two tailed test

If it falls within these limits I am inclined not to reject the null hypothesis, on the grounds that there is a high probability that the difference between the means is due to chance. At the one per cent level of significance I am bound to reject the alternative hypothesis if the mean falls within the range where 99 per cent of the sample means are expected if they belong to the same population.

What is most likely to be my hypothesis in this case? Surely that reading this book will improve a student's chances of a high score in a statistics examination, because it would be self defeating to sales (and to my ego) if I was to suggest that it could actually worsen a reader's performance! Thus my alternative hypothesis is going to be more restrictive than an alternative to the null hypothesis that allows for either of the outcomes at the one or other of the tails: I hypothesize that a reader's performance will be significantly better (that is, I consider the right hand tail only).

In test terms we would require the difference in performance between readers and non-readers to be so large that only something like a one per cent chance existed that it was the result of sampling and not a real difference in abilities. But note that this one per cent chance is now confined to a single tail of the normal distribution, that is, the right-hand tail, and it is not divided between both tails. We are saying in effect that there is a 99 per cent range of acceptable sampling errors that would allow the mean of the readers' scores to fall to the left of the 1 per cent range at the right-hand tail and still allow us to believe that both readers and non-readers were in the same population (figure 11.2).

If, however, the mean falls to the right of that 99 per cent range, then we have reason to believe, with a 99 per cent probability, that the readers come from a different population to non-readers, that is, that reading this book does make a significant difference in their examinations. The calculation used in these one-tailed tests requires access to a Z-table (usually published in statistics textbooks as an appendix) and it need not detain us here.

There is one small issue left to consider. If we require that the readers' mean score is significantly higher than the mean of

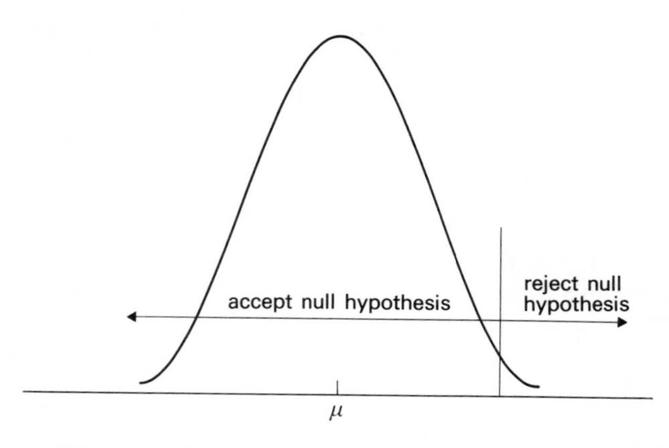

Figure 11.2 Decision rule for a one tailed test

non-readers we are in one sense making it easier to establish that there is a significant difference in the two groups, and therefore, easier to reject the null hypothesis. If the difference in the sample mean can fall within one per cent of the area under the right hand tail and still be considered to be from a different population (that is, reading this book does make a difference to examination results) it must be an easier target to reach than when a two tail test is tried, if only because the 1 per cent level of significance is divided between each tail to ensure a 99 per cent probability of any difference being due to sampling error.

This must increase the probability of a Type I error – claiming that reading this book is significant when it is not (and also ignoring the possibility from the other tail that reading this book is positively harmful!).

Chi-squares

You will not read very far in sociology, marketing, psychology and industrial relations without being struck by the qualitative statistical treatments that are commonly used in these subjects. These are in contrast to the quantitative treatments found in economics. The differences correspond to what statisticians call *parametric* and *non-parametric* statistics.

In parametric statistics we are concerned strictly with the parameters of a population (its mean, standard deviation and distribution), and in non-parametric statistics we are concerned with the attributes of a population (gender, possession or non-possession of a characteristic, broadly grouped categories like 'old', 'young', 'short', 'tall', 'agree', 'disagree', 'like', dislike' and so on) rather than with variables (age, height, weight, length, duration, scores and so on).

In social science there is great scope for non-parametric statistics both from the point of view of the availablility of suitable classificatory data and also in the simplicity of its mathematical computation. Among the most common statistical tests used in the subjects mentioned above, the *chi-square* test (chi is pronounced 'kye' as in 'sky') must be a favourite one. It was developed by Karl Pearson in 1900, and received some constructive criticism from Fisher in 1922 – though Pearson did not see it that way and it was for many years a source of conflict between them (Fisher's view eventually prevailed).

The gist of the chi-square test is to compare the *a priori* theoretically *expected* frequency with the *observed* frequency of particular attributes in some data. If the discrepancy between these frequencies is above a given level (calculated from a chi-square table found in all textbooks) then a probability of a particular relationship between the attributes is suggested and the null hypothesis is rejected; if the discrepancy is below a given level then the alternative hypothesis is rejected.

The method is similar therefore to that for using t and Z tests and setting the null against the alternative hypothesis. The only difference is in the nature of the data to which it can be applied. Suppose we wanted to test if there was a relationship between the sex gender and subject choice in a sixth form, or between sex gender and occupational intentions of a group of university students, or similar such relationships.

If there was no difference in the consequence of being of one sex as opposed to another we would expect the same proportions of each sex to make the same subject or occupational choice. The null hypothesis would be that there was no association between the attributes, and the alternative hypothesis would be that there was an association.

Basically the calculation consists of summing the differences between each observed value with its expected value (the formula for this can be found in a textbook) and dividing this by the expected value. Roughly, if there is a large value for the chi-square in your calculation this suggests an association between the attributes; conversely for a small chi-square value. How big is 'large' and how small is 'small' is decided by comparing the calculated value with those in the chi-square table.

The chi-square table shows the pre-calculated values for varying degrees of freedom that are associated with the number of attributes that are being compared. A 2 × 2 matrix (two attributes associated with two other attributes, such as, male and female and, say, car drivers or non-car drivers) has one degree of freedom; a 3 × 2 matrix has two degrees of freedom and so on; the calculation of the degrees of freedom is extremely simple and uses a simple formula.

For eight degrees of freedom, the chi-square table shows that the chi-square you calculate must be greater than 15.5 for there to be a 5 per cent chance that the null hypothesis of no association between the attributes is true. For a 1 per cent chance of the null hypothesis being true, the chi-square has to be greater than 20.1.

Suppose that your chi-square calculation produces a number such as 23.7. You might decide on that basis that the probability of there being no association between the attributes is too remote (under 1 per cent) and you would therefore be inclined to reject the null hypothesis. The alternative hypothesis that there is an association is significant at the 1 per cent level (23.7 is larger than 20.1).

Chi-square testing can be used for a wide variety of cases and it is used widely in the social sciences. Providing certain basic conditions apply, there is good reason to believe that chi-square tests are appropriate. As with most significance tests, chi-square requires that the sample observations are independent of each other (otherwise spurious association may occur), and that they are collected in a random manner. It is also essential that there are at least 50 observations and that (normally) there are at least five observations in each category (for example, if we are testing for association between sex gender and car ownership we would need at least five instances for each possibility, otherwise the difference between the observed and the expected frequency would be too large before the calculation of the chi-square value was made).

Correlation between variables

Francis Galton discovered the significance of the relationship between variables as a result of many years of painstaking collection of data on inheritance. He plotted the corresponding values of certain variables for parents and their offspring (Pearson, E. S., 1967) and noted that the means of the variables were roughly linear (see figure 11.3 for a reproduction of his original graph). In this he discovered what has come to be

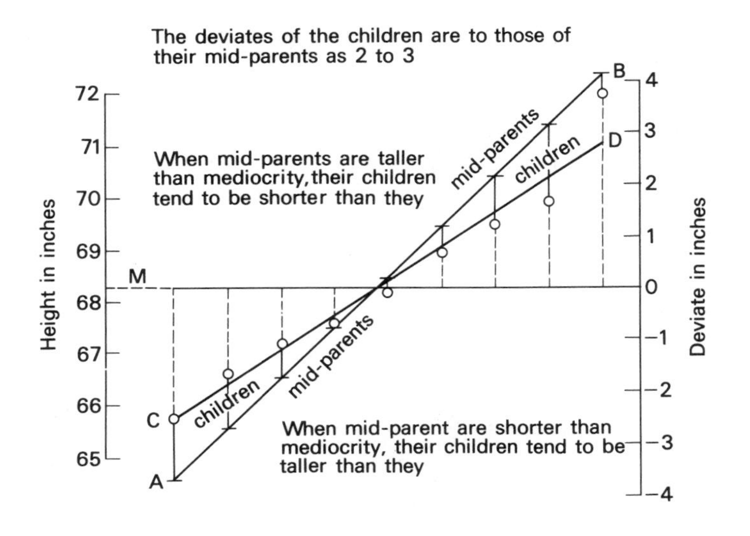

Figure 11.3 Galton's regression diagram for hereditary stature (1885)

known as *regression,* and drew several conclusions from this in his research into heridity.

Galton's most important discovery came in one of those flashes of insight that surge in the mind, almost without invitation, when we are working on a problem. On a railway platform near Ramsgate he was pondering his data when he suddenly perceived a regularity in his results: the variables from his samples were distributed about a straight line in a regular pattern.

If, he noted, the values of the variables were linked together according to their value the result was a broadly ellipitical shape with the regression line as an axis. In this leap of the mind, correlation was born (though earlier writers had stumbled across its existence without elaborating its mathematics) (see figure 11.4).

A Cambridge mathematician, Hamilton Dickson, provided

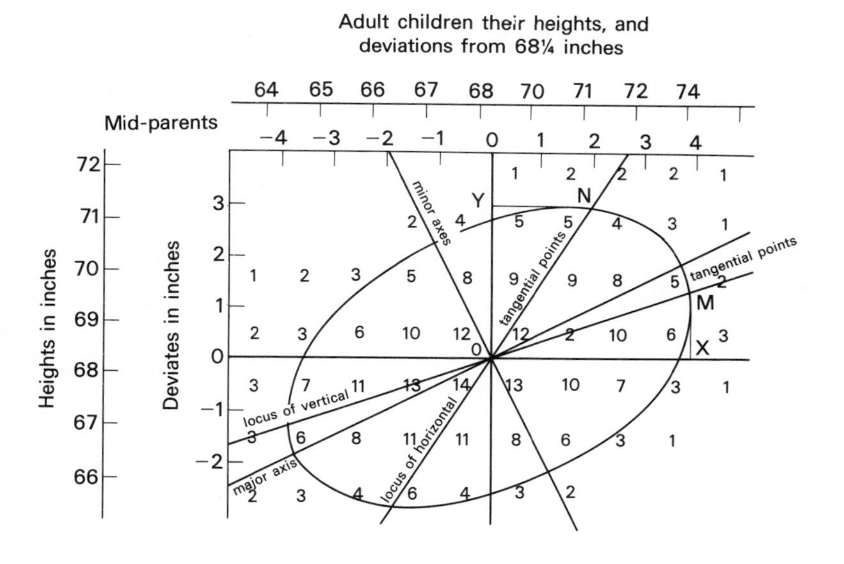

Figure 11.4 Galton's elliptical contour diagram (1885)

Galton with the mathematical insight into the properties exhibited by his data, and two years later, in 1888, the correlation coefficient saw the light of day for the first time. It soon became one of the most common features of statistical analysis and plays a major role in the latter stages of most statistics courses.

Briefly, correlation and regression analysis are used to estimate the *strength* of a particular relationship between two variables and to make *predictions* about the possible relationship for different values of the variables.

The mathematics of correlation and regression look a lot more fearsome than they are in fact, and the basic ideas behind them are quite simple. Suppose we want to know whether the marks students get in mathematics are an indicator of the marks they would get in economics. This is a problem for correlation and regression analysis. One way to see if such a relationship exists is to plot the corresponding marks in each subject on graph paper, using the vertical axis for each student's mark in mathematics and the horizontal axis for the corresponding mark each student got in economics.

There are three basic possibilities that we can observe for the general pattern of the entries in such a graph. The pattern might conform roughly to a straight line rising upwards from left to right. This suggests (though does not prove) that the mark in mathematics is a *positive* indicator of the mark in economics; students who are 'good' at mathematics tend to be 'good' at economics (positive correlation, see figure 11.5).

The straight line does not have to be upward sloping in this way, for it could be downward sloping from left to right, indicating that there is an *inverse* relationship: a high mark in mathematics is associated with a low mark in economics; students who are 'good' at mathematics are 'poor' at economics (negative correlation, figure 11.6).

Finally, the scatter of points entered on the graph could be irregular, such that no association between performance in the two subjects exists; a person doing well in mathematics may or may not do well in economics (zero correlation, figure 11.7).

Using a graph in this way is often a first step in an analysis – it presents the data in a visual form and at a glance produces

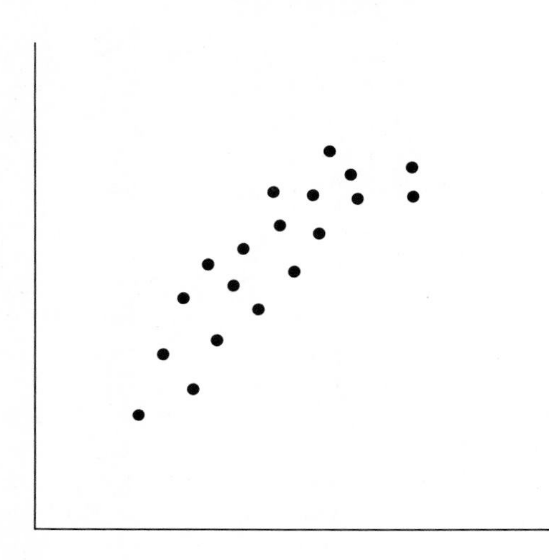

Figure 11.5 Positive correlation

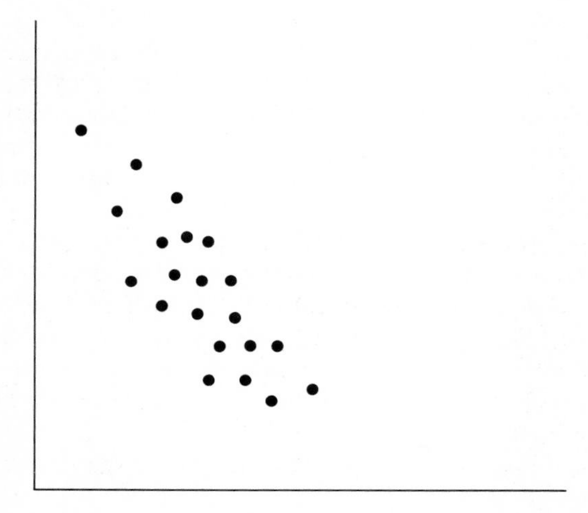

Figure 11.6 Negative correlation

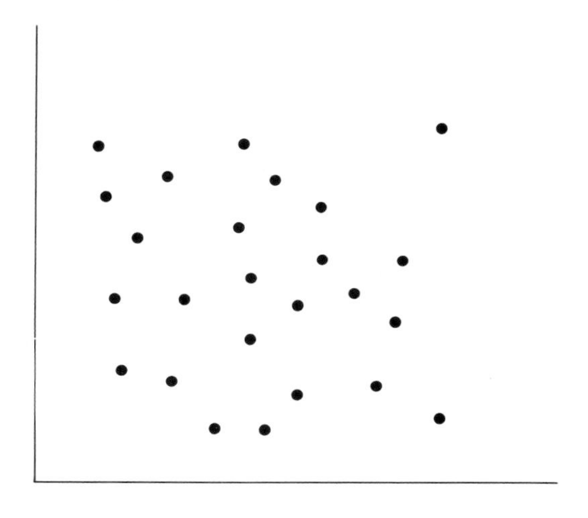

Figure 11.7 Zero correlation

evidence for or against a hypothetical relationship. Its simplicity is an advantage because much ground can be covered quickly by busy researchers if they can eliminate supposed relationships with an economy of effort by using a scatter diagram.

Clearly, the evidence of the *scatter diagram* is not sufficient in itself to establish whether a real relationship exists for the relationship may indeed be spurious (the migration of storks corresponding to a surge in the birth rate does not 'prove' babies are brought by storks). Nor does a scatter diagram show the extent of the intensity of a relationship if one appears to exist. For that we need to calculate the *correlation coefficient.*

Demonstrating the calculation of a correlation coefficent would impose too much manipulative material on you for me to meet my objective of separating out the mathematics from the underlying statistical principles of the techniques we are discussing. The gist of the coefficient is a numerical value for the strength of the relationship between one variable and another. One of the most common measures (there are in fact several different ones) is Pearson's *product moment correlation coefficient,* which measures how well the data fits a straight line (it is useless for relationships of a non-linear character).

The coefficient will take a value running from –1 to +1, with 0 indicating no correlation at all. Values that are negative suggest a negative correlation (that is, as one variable increases in value the other diminishes), and values that are positive suggest a positive correlation (as one variable increases the other increases).

We have exactly the same situation in correlation analysis as we have with any other statistical sample: how sure can we be that the observed relationship is valid for the population on the basis of a sample from it? The answer is found by a method that is analagous to those used in the significance tests discussed earlier.

We set out a null hypothesis that the population correlation coefficient r is zero and the alternative hypothesis that it is not zero. We use a simple formula to give us a value which we can compare with the 'Student's t' table to establish a degree of belief that the sample correlation coefficient is a reliable estimate of the population correlation coefficient between the two variables.

Alternatively, we can calculate the standard error of the correlation coefficient by a manipulation that consists of squaring the correlation coefficient found for the sample, taking it away from one and dividing the result by the square root of the number of pairs in the sample.

There are two decisive influences on the result: first, the larger the correlation coefficient the larger the number we take from 1 and, therefore, the smaller the number that is divided by the square root of the sample size, and consequently the smaller the standard error. Secondly, the larger the sample size the larger the number we divide into $(1-r^2)$ and the smaller the standard error. It follows that the smaller the standard error the more confidence we have that there is a relationship of some sort between the variables because we either have a high correlation or a large sample (or indeed both).

If we assume that the sample correlation coefficients are normally distributed about the population correlation coefficients, we can make estimates of the likely range within which the population value for r will fall. A normal distribution suggests that we would expect with 68 per cent certainty that the population r is within a range of plus or minus one standard error of the sample r (for a 99 per cent certainty we would expect the population r to be within plus or minus 2.5 times the standard error).

It is essential that you do not fall into the trap of thinking that a correlation coefficient establishes a *causal* relationship between two variables. All it establishes is that there is a relationship (accidental perhaps) without in any way implying that the relationship matters in the slightest degree. It may or may not matter, and the causal relationship may be from variable A to variable B, or from variable B to variable A.

To decide these issues you will need a great deal more information, including perhaps a notional explanation of how the causal mechanism works. The last will require a moderate dosage of elementary common sense if spurious correlations are to be avoided.

Regression

Regression techniques are closely related to correlation and are generally taught adjacent to each other (though the order varies according to taste, teacher, and textbook). The basic idea is simple: the closer two variables are related to each other the more they conform to a straight line when graphed.

Galton was the first to use regression in his research into hereditary relations between parents and their children. He found, for instance, that the height of a parent influenced the height of a child, that is, that tall parents tended to have tall children. The relationship was not perfect because the children's heights tended to be closer to the average height than their parents. In other words, a tall father would produce a tall son but the son would be slightly less tall than the father. Similarly, a small father produced a small son but the son would be taller than the father. Galton called this *regressing to the mean,* and he applied it to other characteristics too (see figure 11.3).

The values of variables, particularly in the social sciences, do not conform to a perfect correlation, as you will soon discover as you plot your research data in scatter diagrams, even where there is a definite relationship between the variables.

If we were to hypothesize that the parents' income strongly influenced the amount that was spent on educating their children we could not insist that this was an exact relationship: some parents may have the same income, but it is unlikely that they would spend exactly the same amount to the nearest pound note on educating their children; from roughly the same income some would spend more and some less, but we would expect (if our hypothesis is true) that a parent with twice the income of another would spend more than the poorer parent, and a parent twice as rich as the richer parent would spend yet more, and so on. How much more was spent in each case is not as important as the fact (if it is a fact) that it is distinctly more.

We would expect to find exact relationships between relevant variables in the physical sciences (the expansion of gases per degree of temperature for instance; the conversion of Centigrade to Fahrenheit temperature scales and so on,) than in the social sciences, if only because it is possible to isolate the relationship between physical variables under laboratory conditions but it is extremely difficult to identify relationships in the social sciences.

Let us assume that the cluster of dots is roughly upwards sloping (indicating a positive correlation). The question posed by regression analysis is: how closely do they conform to a straight line, and, more particularly, which straight line do they most closely correspond to?

The problem boils down to finding the straight line that *best fits* the data. With linear regression we assume that the relationship between the variables is a linear one and we are compelled to this assumption precisely because of the weakness of our hypothesized theoretical relationship. Given all the multifarious factors that operate on and influence variables in society we are, if you like, looking for relationships through a dark glass. We think we can spot something going on, but we are not sure, when we suggest that there is some connection between the two variables. The linear relationship is the easiest to conceive of and to search for, especially as we have a paucity of data; to seek to establish non-linear relationships is a very ambitious research programme indeed. If the relationship is non-linear, and we cannot separate out what form it takes, our linear relationship is going to be a very poor approximation of the situation. It might be so poor that we dismiss the possibility of any relationship at all and make a Type II error (we accept the null hypothesis of no relationship when in fact there is one).

Now the manipulation required to find the regression line looks fearsome at first glance and therefore best left out of our discussion. It boils down to finding the line that minimizes the sum of the squares of the deviations of each dot from a straight line (see figure 11.8). The reason for looking at the sum of the squares of the deviations of the plotted dots is based on the fact that some dots will be above the line and some below it and there is a danger that if we simply summed the absolute deviations they would cancel out and produce a zero.

For a given set of data only one line will minimize the sum of the squares, and the manipulation provides a procedure for

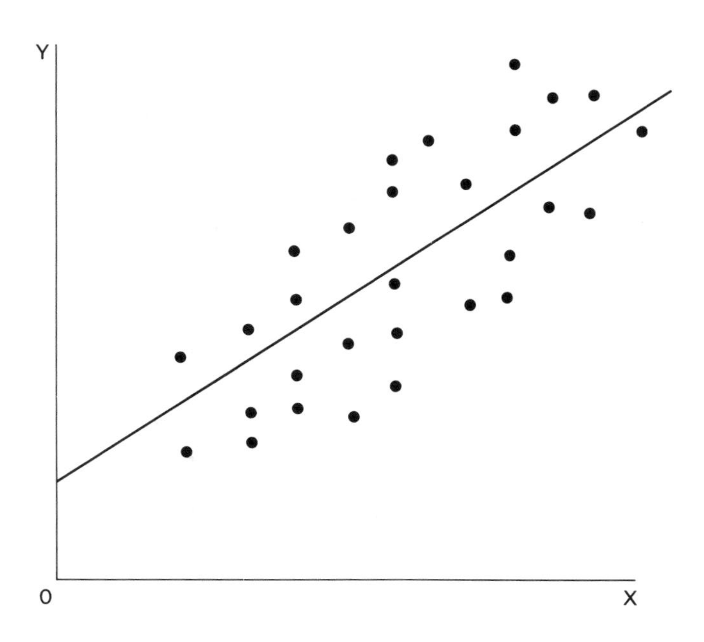

Figure 11.8 Regression line

finding it (in technical parlance it is known as the method of *ordinary least squares,* or OLS). Fortunately, pocket calculators can cope with the arithmetic that is required.

The straight line produced enables us to interpolate within the range of the data we have available, subject of course to error because the line is the one that best fits the data and the values are unlikely to fall exactly on it but will be spread at varying distances about it.

A straight line has two components when written in equation form: the slope and the intercept. The slope gives the rate at which the dependent variable changes for changes in the independent variable. Thus, if there is a linear relationship between time spent studying and our examination marks (can you see which is the dependent and which the independent variable here?), the slope of the regression line tells us by how much our examination marks will increase for each additional hour of study. If the straight line equation was such that $Y = 25 + 2X$, we are saying that for every hour of study we can gain 2 marks to add to the 25 we would get if we sat the examination 'cold' with no studying (see figure 11.9).

Exercise 11.2

How long need we study to get the pass mark of 50?; how many marks would we get for 25 hours study? (see Appendix).

The intercept is simply the 25 marks we get for no studying at all and is shown where the straight line joins the vertical axis. If we got 10 marks without studying, it would join the axis at 10.

Now with social science data we do not have an exact relationship and therefore the equation of the regression line shows the average, or approximate, relationship between studying and examination marks. The plotted dots vary about the straight line and it may be that for a given amount of studying we could obtain slightly more, or slightly less, examination marks than the line predicts.

The extent to which the scatter of points lie closer to the straight line, the more accurately the regression equation predicts, or explains, the relationship. The measure of this

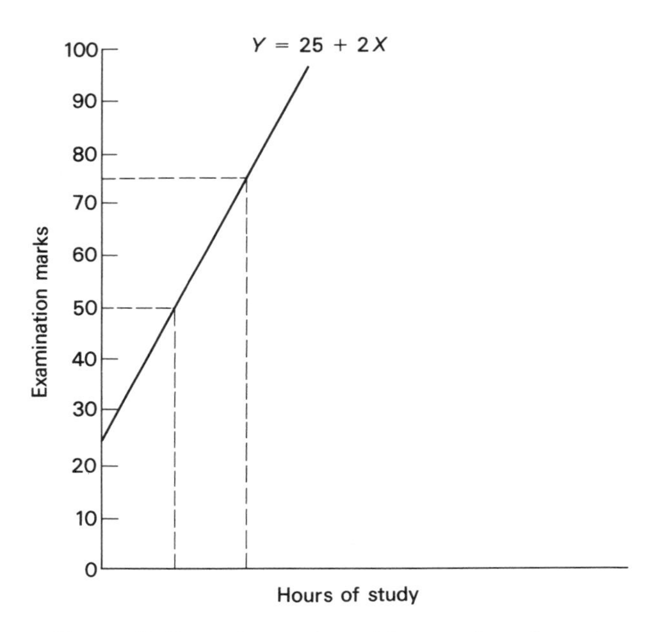

Figure 11.9 Regression of hours of study and examination performance

goodness of fit is known as the coefficient of determination, usually symbolized by R^2. In figure 11.10 a regression line is fitted to the data and we want to explore just how well this fitted line explains the alleged relationship.

If X is the independent variable then for any given X we can read off the Y value. Out of a set of readings relating X to Y the relationship for a given X value could crudely be given as the average of all the Y values ($Y^{\#}$ in figure 11.10). We can improve on that estimate by finding the value of the Y from the regression line at the given X value (Y' in figure 11.10). In this case Y' is less than the actual Y^* value for X in the scatter of dots. This gives us two deviations: first, we have the amount by which the Y value on the line deviates from the average value, that is, the explained variation, and second, we have the amount by which the actual reading Y^* deviates from the line, that is, the unexplained deviation.

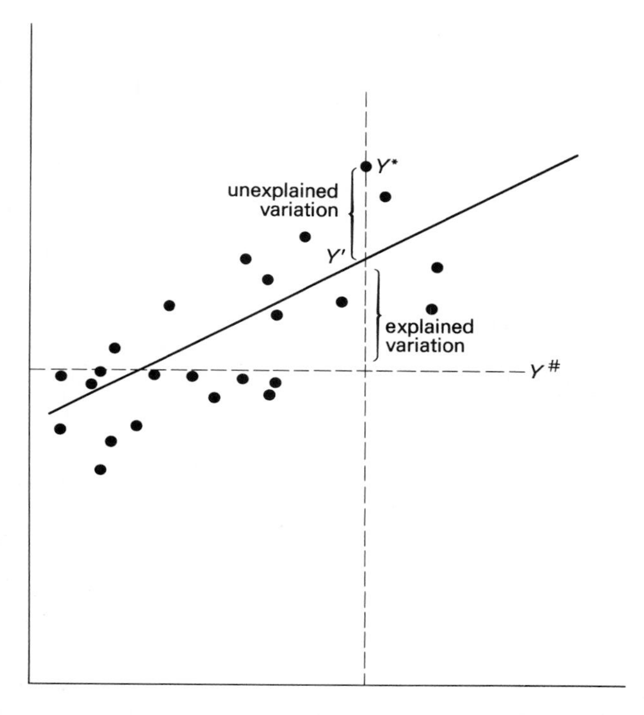

Figure 11.10 *Explained and unexplained variation*

The greater the proportion that the explained deviation ($Y' - Y^{\#}$) is of the total deviation ($Y^* - Y^{\#}$), the more accurate the explanatory power of the regression relationship. (The calculation of R^2 involves finding the ratio of the squares of the explained deviation to the total deviation.) The R^2 ratio can vary from zero (no explanatory power) through to one (perfect explanatory power), and normally lies between these two extremes. Thus, if the R^2 value is 0.68, we are claiming that the independent variable accounts for 68 per cent of the variation in the dependent or Y variable. It follows that the higher the R^2 value (that is, the closer it is to one) the more confidence we have in the alleged relationship, and conversely. A low R^2 may mean that there is no relationship (conclusion: accept the null hypothesis) or that it is not best represented by a linear regression equation (conclusion: try a non-linear relationship or reject the null hypothesis). You can see that there is a lot of scope for disputes in interpreting statistical data!

What connection is there between the R^2 value and the correlation coefficient, r? Briefly, R^2 is measuring the amount of explanatory power that a knowledge of the independent variable gives for the value of the dependent variable, while r is measuring the statistical correlation between them. An r of 0.6 does not explain the relationship – it merely indicates that for some reason, or by accident, the two variables are correlated. The R^2 for $r = 0.6$ is 0.36, suggesting a much smaller influence of one variable on the other than might be implied by 0.6. Even an $r = 0.9$, which implies a strong correlation, is weakened when the R^2 value is 0.81. In both cases, however, the existence of a causal relationship is not established by the fact of correlation nor the goodness of fit of the regression line.

Little is certain in statistical relationships. We are examining data to see what potential there is for identifying a relationship, but having found that there might be some relationship there we cannot automatically assume that the relationship is definite, causal, and robust. It may be, and it may not be, and that is the irony of statistics: without examining the data we can only guess at what is going on, but in examining the data we may still be left guessing! The difference is, of course, that a guess without any consideration of the data is boundlessly extravagant – choose your prejudice and then fit your guess to it – while, on the other hand, a guess after due consideration of the data is not boundless, it is restricted – our prejudices cannot run directly counter to what the data shows, but the data cannot show our prejudices to be true even if it appears to support them.

12

Statistics and Statisticians

My invitation to statistics is drawing to a close, but before we call it a day, we shall have a last general look at the subject and those who practise it, if only to convince you that it is worth your while to think about studying statistics in greater depth. So, in this final chapter, we shall stand back a little and look at the broader reaches of the canvas we have been close to in the previous chapters and address ourselves, not so much to technical questions, as to what is it all about and why might it be important.

Popular views of statistics

There are those who decry statistics in any shape or form whatsoever. If they had their way, no doubt, they would have John Graunt buried along with the dead of the Mortality Bills. That statistics are unpopular among large numbers of people – and not just those who have had to churn their way through endless number crunching as part of their training – is a fact that those of us who wish the subject well have to accept.

Statistics has a bad press; it is a subject that attracts disdain from otherwise sensible citizens; it does have a bad reputation ('lies, damn'd lies, and statistics'), and it can be used by the guardians of integrity as well as abused by the enemies of the truth. All this is both true and, in a certain sense, irrelevant.

Statistics is not going to be disinvented just because of its popular (or rather, unpopular) image. Statistics will remain an active part of modern life precisely because it only systematizes what most people do all the time – everybody acts on the basis of their beliefs about what is likely to happen to them (insurance is an obvious example). True, in the main, these predictions are often unscientific, superstitious even, and would not stand up to the most elementary test from anybody trained (even badly) in statistics.

Abolish statistics and you would have to censor parents discussing the sex of their unborn child: 'It is going to be a boy, because the first born in my family for three generations has been a boy' and so on, (can you see now what is wrong with that statement?).

You might have to banish the weather as a subject for discussion in every home, office, factory, bus queue and railway carriage (not to mention breakfast TV show, radio programme and evening TV forecast): 'take your raincoat with you today as we predict it will rain in the afternoon', and 'red sky at night, shepherds' delight, red sky in the morning, shepherds' warning'.

Much of the nonsense that passes for statistical truths is quite harmless; it only affects the decisions of those who are misled by what they think is something called the 'law of averages'. If they want to act in the way they do there is probably no harm to the rest of us if we let them get on with it.

A casino was once saved from certain ruin by a misunderstanding of statistical relationships by the gamblers present, who saw the evens come up eight times on the roulette wheel and kept betting that the next spin would produce odds 'because the law of averages says an odd must come up sooner or later'. They bust themselves piling on their bets every spin until they had no money left. After 28 spins the ball fell into an odd number, but the bank was safe as it had all the punters' money, and they could only watch in disgust. It was no fluke. It was an honest game. It is just that the so-called law of averages does not say anything like they thought it did. If they had bet on evens from the start they would have won the casino itself after a few spins (but they did not and their folly joined the long list of gamblers' weepy 'if only' stories).

There are other misinterpretations of statistical probability, some of them laughable and others just plain silly. Coins do not have a memory and when they are tossed in the air they do not consider how they landed the previous time and thus switch from tails to heads. The chances of heads or tails remain the same for every toss (about even); if you do not believe that you will end up bust long before the bank. Likewise, with aircraft: just because a plane crashes in New York it does not mean that it is any safer, or any riskier, to fly that day out of New York, or anywhere else, on the (spurious) grounds that as 'statistically a crash was due and now that it has happened it will not happen again for some time'.

The chances of getting a royal flush in poker are exactly the same as getting any other *named* hand, such as the jack of diamonds, the two of clubs, the eight of hearts, the six of spades and the nine of clubs. True, this might make the game less interesting, if only because only some hands have an obvious title, and also it might be more dangerous, because most players would be extremely suspicious if you named a hand like the one above and then got it!

Lightning does not have a memory, hence it is not true that it does not strike in the same place twice (if it was, then after being struck by lightning, buildings could have their lightning conductors removed!), nor is it true that the safest place for your head on a warship was in the hole just made by a canon ball (any part of the ship is as likely to be hit as any other – the balls, at least the eighteenth-century variety, did not know to divert themselves from parts they had already hit!).

Some people are put off from doing something – flying as opposed to driving – on the grounds of something that happened to someone they know or read about. It may be that many more people are at risk on a motorway than are at risk in an aircraft, but if you believe that flying is more dangerous – even though only 50 people died in plane crashes, say, in a given year, against 5,000 who died in road accidents – then you will not fly and that is that.

We could go on correcting statistical probability howlers that drop into everyday conversations and in many cases actually guide the actions of individuals but we shall move on instead.

Spurious statistics

There are so many ways in which statistical data can be misused to present misleading suppositions that whole books have been written that identify the most common dodges and pitfalls. At the very least, a short exposure to statistical methods would caution you against accepting what appear to be plausible figures, often put forward with a straight face by people who have causes to defend or grievances to air.

Just to note a few of these, we can think of the number of times in which arguments (indeed strikes) have been fought over percentage wage increases as if a specific percentage had the same meaning to all groups of employees. A university lecturer complaining about a 4 per cent pay award (on an average salary in excess of £12,000) might have less to complain about than a school janitor who has a claim in for 8 per cent (on an average salary close to £5,000). To accuse the janitors of exceeding the inflation rate and endangering the economy might appear to be a trifle far-fetched when the percentage increases are translated into actual money terms.

When you see percentages you should wonder what they are percentages of and not automatically rank one against the other by their relative size. A poll that shows 30 per cent in favour of something must be translated into 30 per cent of the number polled. This will look less convincing if only nine persons are interviewed, compared to nine thousand (and even less convincing if those polled were chosen by other than a random method).

The most obvious caveat needed when dealing with statistics concerns the alleged causation occurring when two variables appear to vary in tandem with each other. Now sometimes the case for a casual relationship can be made on the basis of the associated evidence – the sun rises during the day and so does the temperature – but in many cases the relationship, while attractive to the instinct and ideas of the researcher, may be utterly spurious.

Statistics texts remind us time and time again: the fact of a correlation between two variables is not a proof that there is a causal relationship, and if a causal relationship appears possible it is not necessarily obvious in which direction the causality runs. Understanding this most important caveat is a sign of your appreciation of the statistical method.

Does smoking tobacco affect the chances of a person developing cancers? The answer is not at all straightforward, at least according to some people. Fisher, for example, shocked a lot of his colleagues – they still argue about it – because he examined the evidence for the relationship between smoking and cancers and stated that it had not been established *statistically* that there was a relationship. Whether you accept that conclusion (and I confess I do not) is beside the point; what counts is that the statistical relationship is not obvious just because sets of data suggest that there is some connection.

The data has to be backed by other evidence – perhaps an explanation of how smoking influences the incidence of cancer – and this evidence itself must be open to debate and enquiry. The spurious relationship between the migration of storks and the birth rate is open to challenge because there are alternative explanations of how babies are conceived and born and these explanations do not involve the activities of storks.

Much the same applies when we consider the question of tests of significance. It cannot be stressed too often that just because a test shows something to be significant in a statistical sense, this does not mean at all that it necessarily is important. It may or may not be important. There may be something lacking in the research design and this may conceal important factors as much as reveal unimportant factors.

Statistics, in other words, is not a method divorced from other evidence. If a relationship appears to be present it is not enough to sit back and say that this is the case: the scientist is expected to produce an explanation for the connection and cannot merely appeal to the evidence of data alone.

To rely entirely on a set of numbers without a theoretical explanation for them is a highly risky strategy, especially if you are concerned about your reputation. If the relationship that you allege is shown on other grounds to be spurious, or is shown to have another explanation (perhaps the reverse of your inferences), at best you might have a debate on your hands, at worst you might be made to look silly. Hence, as a user of statistics you have to learn to distinguish between plausible assertion and possible evidence, both in respect of your own work and the work of others.

What makes a good statistician?

Statistics is both a discipline in its own right and a set of techniques at the disposal of other disciplines. Professional statisticians are somewhat proud of their calling and a trifle testy with those who, though not fully qualified in statistics, nevertheless have pretentions to knowledge about the subject.

To some extent this is justified and reflects the immense progress made in statistical methods over the past forty years. Statistics has definitely conquered the social sciences and few can claim any authority in their field who have not gained some understanding of the statistical method. Those who work with statistical ideas and manipulations, even at the very low level of sophistication as represented in this book, are way ahead in seriousness compared to those who shut their eyes to any forms of evidence – in case it is contrary to their own views – and to those to whom statistics is an absolutely closed book.

The fact that specialists in statistics are operating at a frontier not much divorced from advanced mathematics does not make them more worthy of respect on this ground alone. They have created – and continue to create – the tools that other people put to good or ill use, but as to which is primary, the creator or the persons who apply that which is created, is something nobody can answer with any certainty; you choose your heroes and I will choose mine.

While we cannot all aspire to grasp the inner essentials of the abstract ideas and concepts of statistical theory, we can, in large measure, grasp the bare essentials of the meaning of particular ideas and concepts, as well as their uses and limitations.

Data is not something sacrosanct. It has a purpose, but at the end of the day it is only as good as your skills in assembling, manipulating and presenting it. Inference is a weak reed if the data base is of exceptionally poor quality and without substance. With a little practice almost anybody can think statistically. Even the most elementary of exposure to statistical methods gives the user something to think about when confronted with claims and assertions about aspects of the real world.

Is the data sound? This is an obvious first question to put. Where did the data come from, who collected it, how consistent is it, how representative is it, how justified are the categories it is organized into, what has been excluded and why, what deficiences are there, and what has the reporter said about them? These must spring to mind at a glance at the first page of data brought before you.

Statisticians read the small print at the bottom of a table. They look for the exceptions and the exclusions. They question the breaks in the series. They want to know why the author has jumped across categories and time periods. In short, they examine the material closely.

When they look at the results of manipulations they look at the properties of the significance tests that the author has applied – and they note if some obvious tests appear not to have been reported: does this mean they were not tried or were they tried and excluded because the results proved embarrassing? Given the opportunity they will run their own tests on the data and check the arithmetic; they might even apply some different tests just in case they prove to be interesting. In other words they test each other's work to see how robust it is.

They are also interested in the overall research design of any project they are studying. The data used may indeed be magnificent and the manipulations may be precise and accurate, yet the phenomenon they claim to be measuring may not be best measured in the way they have proceeded. This might be apparent in the research design (often described in projects and dissertations as the research *methodology,* though this is in fact a wrong use of the word methodology), and statisticians are only doing their job when they question whether the evidence actually is relevant to the phenomenon it is allegedly explaining.

It is up to the researcher to present and defend his or her design and much time and trouble would be saved if this was done properly to start with. If an experiment does not relate to a problem in the way it is claimed, it follows that the evidence adduced to support the research into the problem is redundant. It is no good testing the attitudes of the nation to gambling by asking the opinions of those in the queue for the train to the Newmarket races. Anything discovered in such a biased sample is worthless if it is presented as being the views of a random sample of the population at large.

In summary, a statistician is concerned with the design of the research (does it do the job claimed for it?), the quality of the data (is it accurate?), and the manipulation of the data (is it an appropriate test of the stated hypotheses?).

Now not all of us have the time or the inclination to acquire the many skills in these areas that distinguish a professionally trained statistician from the rest of us, but we can appreciate something by exposure to the basic principles of statistics. That is why reputable institutions insist that their graduates take some course in statistical methods, if only at the appreciation rather than the qualifying level.

Jobs for statisticians?

Employment prospects for persons who have some knowledge of statistics are bound to be influenced by the state of the economy. If, however, it was left purely to the demand for the output of statisticians I think there is likely to be an increasing, rather than a decreasing, demand for their services.

The flow of statistical data appears as an increasing function of the complexity of society: government departments appear to demand more and more data from companies (some of their representatives are to be heard from time to time complaining about the official returns they are now required to make under one piece of legislation or another), and much of this data reappears later in the form of statistical tables and summary reports.

If the statistical work of multi-governmental institutions such as the European Economic Community, the United Nations and its various agencies, and the numerous international bodies that have proliferated since the Second World War, is also considered, it would seem that a career as a statistician must have some element of job security in it that might perhaps be missing in other occupations. This is particularly true when it is remembered that wherever decisions are likely to be made on the basis of data collected by a central agency, then there is an automatic requirement that the people likely to be affected by those decisions acquire the services of interpreters of the data, if only to protect and enhance their own interests. Thus trade unions, employers' associations, professional associations, other governmental and quasi-governmental agencies, political parties, pressure groups, and public relations organizations, all have a need for statistical competence either in the form of specialists trained in statistics or in the form of people with other talents but who also have some technical competence in statistical methods.

Many wealth creating activities require statistics for their everyday functioning. Manufacturing companies use statistical methods to ensure and maintain quality control in their inputs and their products. They cannot test every item to destruction and therefore can benefit from well chosen samples and safe quality limits. The marketing function has to use statistical methods to develop marketing plans, to estimate demand for their products, and to test their pricing strategies. There is a limit to the amount of actual market experimentation they can indulge in before they use up irreplaceable resources and go bust. Financial bodies use and process vast amounts of data, and stockbrokers spend a great deal of time watching for shifts in share and bond prices and anticipating shifts in interest rates and so on. They could not even begin this work without a knowledge of statistical techniques.

I do not want to exaggerate the situation: a knowlege of statistics is not an automatic meal ticket for life (would that it was, would that it was!), but there is something to be said for the view that a knowledge of statistics will not harm you, and might positively help you, in the career you intend to pursue.

Even as a citizen, a knowledge of statistics is of some use - after all, opinion formers permanently inundate you with statistical concepts of more or less doubtful provenance and you have to be able judge the reliability of what they are telling you if you wish to make informed decisions about yourself and your prospects.

In academic research, especially in the social sciences, a knowledge of statistics is of such obvious importance that it is sometimes a small wonder that it is still discussed as a 'problem' from time to time. Bemoaning the relatively poor quality of the statistical basis of some research projects is a hardy perennial in various disciplines. Hence, those who aspire to an academic career, either in research or teaching, ought to accept from the start that they will have to master at least elementary statistical ideas if their career is to progress from aspiration to actuality.

In mitigation

At this point I am about to take my leave of you. This invitation to statistics ends with you, I hope, better placed than you were when you began at chapter 1 to decide how much, or how little, statistics fits into your future plans.

In this invitation I have played down the number-crunching aspects of the subject. This may or may not have been a good idea (my intentions were honourable) and I hope that you do not feel too cheated if the deluge of numbers and formulae associated with elementary statistical manipulations wearies you. Remember, everybody feels the same at their first exposure to number crunching, and normally it wears off sooner or later. With the computer packages that are now available, and ready access to a pocket calculator, you should rapidly graduate from menial tasks to the more interesting interpretative work, once you understand what you are doing, and why.

What you have read may have put you off statistics altogether, and you will not be alone if that is your decision; unfortunately there are many casualities among beginners. I am comforted to some extent by the belief (not statistically tested!) that my invitation will have brought to your attention *something* useful about the subject and that this alone is much better than your having passed it by without even a half-interested glance. If you ever change your mind about statistics, you know now where to start looking.

In the main, however, I hope that you have been made curious enough about statistics to persevere with your initial interests. To this end you might care to consult a short guided bibliography for your next steps towards statistical knowledge.

A final appeal: if (when), in your subsequent explorations of statistics, the going gets a little tough, as anything worthwhile is bound to at one time or another, and you feel that statistics is humourless dry-as-dust subject, you should recall some of the stories of the early statisticians I have included in this book and realise that they were as human as anybody can claim to be. Persevere, for in my view, a social science without the rigour of a common language such as statistics leaves us:

> Here as on a darkling plain
> Swept with confused alarms of struggle and flight
> Where ignorant armies clash by night.

Matthew Arnold: Dover Beach

Further Reading

Basic texts

If you are going to pursue your studies of statistics it is inevitable, and indeed necessary, that you consult a basic textbook in statistical methods and techniques, and the shelves, literally, are groaning with introductory statistics textbooks. Making a selection to suit your needs and interests is something I must leave to you; tastes and preferences cannot be delegated!

The level of difficulty at which you choose to enter the subject will determine to a large extent the type of book you select. In this respect I can mention one or two titles at each end of the spectrum: consult them first, then choose. Two books which have long been favourites at the popular level are:

Darrell Huff, *How to Lie With Statistics,* Pelican, London, 1973,

and

W. J. Reichman, *Uses and Abuses of Statistics,* Penguin, Harmondsworth, 1964.

These will provide many hours of amusement and instruction and are ideal introductions to the darker side of statistical manipulation. Moving on to instruction texts in basic techniques, could I suggest an extremely slim but valuable little text for those who want to begin very gently but who also want to cover a lot of ground swiftly? It was first published in 1966 but is still in print (1982):

G. Kalton, *Introduction to Statistical Ideas: for social scientists,* Chapman and Hall, London, 1966.

Another beginner's text is
Ronald Meek, *Figuring Out Society,* Fontana, London,

which covers much of the statistical method in a most amusing and entertaining way.

Stepping up the level of difficulty, though keeping well within the bounds of readers who have persevered this far, there is:

Dereck Rowntree, *Statistics Without Tears: a primer for non methematicians,* Pelican, London, 1981.

This delivers what it promises: it covers much of the core of the statistical method without even beginner's mathematics. It is extremely well written and is as clear as crystal.

There are many comprehensive textbooks – too many to list here – and as they cover more or less the same ground there is not a lot to choose between them. Among those available I suggest you might try:

John D. Hey, *Statistics in Economics,* Martin Robertson, Oxford, 1974.

R.J. and T.H. Wonnacott, *Introductory Statistics,* John Wiley, New York, 3rd ed. 1977.

K.A. Yeoman, *Statistics for Social Scientists,* Penguin, Harmondsworth, 1968, 2 vols.

On the other hand there is a dearth of critical books. One of these is:

John Irvine, Ian Miles and Jeff Evans (eds.), *Demystifying Social Statistics,* Pluto Press, 1979.

This presents a series of papers on statistics written from a Marxian perspective, but allowing for its advertised views, there are several very useful and interesting papers contained in it, and the two I would recommend that you consult, if only to get another view, are:

Liz Atkins & David Jarrett: 'The significance of "significance tests"', pp. 87–109,

and

Donald MacKenzie 'Eugenics and the Rise of Mathematical Statistics in Britain', pp. 39–50.

(In my view this last is a controversial interpretation of the role of the Eugenics laboratory, but worthy of consideration).

History of Statistics

The history of a subject is always a sound test of its relevance. For a 'good read' consult the fascinating account of:

F.N. David Games, *Gods, and Gambling: a history of probability and statistical ideas,* Griffin, London, 1962.

The best book on dice is:

John Scarne, *Scarne on Dice,* Harrisburg, Pa., 1962.

In addition, two volumes of selected papers in the history of statistics and probability are well worth dipping into and they cover an astonishing amount of material. They are:

E. S. Pearson & M. G. Kendall (eds.), *Studies in the History of Statistics and Probability,* Griffin, London, 1970,

and

M. G. Kendall & R. L. Plackett (eds), *Studies in the History of Statistics and Probability,* (vol.2), Griffin, London, 1977.

I particularly recommend the articles by:

Fitzpatrick, Paul J. 'Leading British statisticians of the nineteenth century', *Journal of the American Statistics Association,* vol. 55, 1960, pp. 38–70; Kendall and Plackett, 1977, pp. 180–212.

Greenwood, Major, 'Medical statistics from Graunt to Farr', *Biometrika,* vol. 32, 1941, pp.101–27; 1942, pp. 203–25; vol. 33, 1942, pp. 1–24; Pearson and Kendall, 1970, pp. 47–120.

Hasofer, A. M.: 'Random mechanisms in Talmudic literature', *Biometrika,* vol. 54, 1967, pp. 316–21; Pearson and Kendall, 1970, pp. 39–43.

Kendall, M. G. 'Where shall the history of statistics begin?' *Biometrika,* vol. 47, 1960, pp. 447–49; Pearson and Kendall, 1970, pp. 45–6
Kopf, E. W. 'Florence Nightingale as a statistician' *Journal of the American Statistics Association,* vol. 15, 1916, pp. 388–404; Kendall and Plackett, 1972, pp. 310–26.
Lazarsfeld, Paul F. 'Notes on the history of quantification in sociology – trends, sources and problems' Isis, vol. 52, 1961, pp. 277–333; Kendall and Plackett, 1972, pp. 213–69.
Pearson, E. S. 'William Sealy Gosset, 1876–1937: "Student" as a statistician' *Biometrika,* vol. 30, 1939, pp. 205–50; Pearson and Kendall, 1970, pp. 360–403.
Pearson, E. S. 'The Neyman–Pearson story: 1926–1934; historical sidelights on an episode in Anglo–Polish collaboration' in *Festschrift for J. Neyman,* New York, 1966.
Pearson, E. S. 'Some incidents in the early history of biometry and statistics, 1890–94' *Biometrika,* vol. 52, 1965, pp. 3–18; Pearson and Kendall, 1970, pp. 323–338.
Pearson, E. S. 'Some reflexions on continuity in the development of mathematical statistics, 1885–1920' *Biometrika,* vol. 54, 1967 pp. 341–55; Pearson and Kendall, 1970, pp. 339–53.
Pearson, Karl 'Walter Frank Raphael Weldon, 1860–1906' *Biometrika,* vol. 5, 1906 pp. 1–52; Pearson and Kendall, 1970, pp. 265–321.
Plackett, R. L. 'The principle of the arithmetic mean' *Biometrika,* vol. 45, 1958, pp. 130–55; Pearson and Kendall, 1970, pp. 121–26.
Rabinovitch, Nachum L. 'Probability in the Talmud', *Biometrika,* vol. 56, 1969, pp. 437–41; Kendall and Plackett, 1977, pp. 15–19.
Rabinovitch, Nachum L. 'Combinations and probability in Rabbinic literature' *Biometrika,* 1970, pp. 203–5; Kendall and Plackett, 1977, pp. 21–3.
Royston, Erica 'A note on the history of the graphical presentation of data' *Biometrika,* vol. 42, 1955, pp. 241–7; Pearson and Kendall, 1970, pp. 173–81.

Another history, by Karl Pearson, though a heavy one, is well worth the effort to browse through, particularly on the life and work of John Graunt and William Petty. It was published long after his death by his son, E. S Pearson. You could spend a happy week consulting it and not waste a minute:

E.S. Pearson (ed.), *The History of Statistics in the 17th and 18th Centuries: against the changing background of intellectual, scientific and religious thought: Lectures by Karl Pearson given at University College London, during the academic sessions 1921–23*, Griffin, London, 1978.

Two more specialized books can be mentioned here and you should have no problem in reading them for pleasure as well as instruction. The first is:

Ian Hacking's: *The Emergence of Probability: a philosophical study of the early ideas about probability, induction and statistical inference,* Cambridge University Press, Cambridge, 1975,

and the other is:

L. E. Maistrov: *Probability Theory: a historical sketch,* (trans. S. Kotz), Academic Press, New York (1967), 1974.

Biographies

By their works ye shall know them, and you can glean a great deal about statisticians and their work from a couple of biographies. The first I would recommend is the biography of Francis Galton:

D. W. Forrest *Francis Galton: The life and work of a Victorian Genius,* Paul Elk, London, 1974.

This has to be one of the best biographies of a scientist for many a year and you will learn a lot about the man and his statistics from reading it.

Somewhat more difficult is a biography of Fisher:

Joan Box, *R. A. Fisher: the life of a scientist,* New York, 1978.

This includes a full bibliography of Fisher's scientific papers. It would be invidious not to mention here the extended biographical essay by J. B. S. Haldane on the life and work of Karl Pearson:

J. B. S. Haldane, 'Karl Pearson, 1857–1957' in E. S. Pearson and Kendall(eds.), *Studies in the History of Statistics and Probability,* Griffin, London, 1970 (also published in *Biometrika,* 44, 1957, pp. 303–13).

Technical topics

For probability theories you can consult several lightly written texts and among the best I suggest the following:

Darrell Huff, *How To Take A Chance: a light hearted introduction to the laws of probability,* Pelican, 1965,

or

Warren Weaver, *Lady Luck: the theory of probability,* Pelican, 1981.

Alfred Renyi has written some 'spoof' letters purporting to be from Blaise Pascal on elementary probability ideas and associated probabilistic philosophy, which are both intructive and brilliantly realistic:

A. Renyi, *Letters on Probability,* (trans. L. Vekerdi),Wayne State University Press, Detriot, 1972.

Moving into the more difficult area, perhaps best left until you are acquainted with more advanced work, there is the authoritative treatise:

John Maynard Keynes, *Treatise on Probability,* Collected Writings, vol. 7, Macmillan, London, (1921) 1973.

A useful little series of individual texts on topics in statistics was published in the 1970s. These are well worth selecting from in order to cover topics at an elementary level. They can be used to support a textbook or to get an overview of a subject. I would recommend working your way through:

Paul E. Spector, *Research Designs,* Sage University Paper Series on Quantitative Applications in the Social Sciences, no. 23, Sage Publications, London, 1981.

Ramon E. Henkel, *Tests of Significance,* Sage University paper (op. cit.), no. 4, London, 1976.

Michael S. Lewis-Beck, *Applied Regression: an introduction,* Sage University Paper (op. cit.), no. 22, London, 1980.

Charles W. Ostrom Jun. *Time Series Analysis Regression techniques,* Sage University Paper (op. cit.), no. 9, London, 1977.

Anybody handling data needs to be reminded, occasionally, of the fragility of its purported accuracy and the best book to summarize this is:

Oskar Morgenstern, *On the Accuracy of Economic Observations,* Princeton University Press, Princeton, New Jersey, 2nd ed., 1963.

Beyond these references we would be entering into a veritable library list of texts, monographs and papers, most either out of reach because they are out of print and only available in good libraries, or unnecessary for my recommendation because if you were likely to consult them you would already be well into a serious statistics course. As my invitation is directed at persuading you to undertake further studies in statistics I will refrain from intimidating you with a too extensive reading list and take my leave of you.

Appendix
Answers and Comments to Exercises

Exercise 3.1

Samples of one are useless as guides to general behaviour or experience. To believe that a single person's experience (in this case as a heavy smoker) is representative of the entire population is a common error. People do generalize in this way, usually from personal experience (for example 'I never had any trouble getting a job so I can't see why the unemployed don't go out and look for work if they want it'). To some extent this is understandable when a single instance is all the experience you have to go on (for example 'I would not eat at that restaurant again – they burned my steak, served chilled claret, and their sweet trolley had flies on it').

When we want to make general statements, however, we must take care not to enhance the importance of a single instance above what it is worth. We must consider the strong possibility that other instances will modify the significance of the single instance.

In the case of a possible connection between smoking and cancer, we would want to test the hypothesis that there was a connection by studying a large random sample of smokers and non-smokers. We could study their medical histories and test for propensities to contract cancers in the two groups. To have confidence in the hypothesis that there is a connection between smoking and cancer we would look for a significant difference in the number of cancer patients in the two groups.

Another method would be to examine the death statistics of the population and see the distribution of cancers among smokers and non-smokers. Statistical techniques, discussed later in the book, look at the techniques for conducting hypothesis testing, and their problems.

Exercise 3.2

The sales by commodity can be shown using a pie diagram by taking the percentage shares for each commodity as percentage shares of the circle. I divided the pie into quarters and used one of the quarters to represent the share going to oil cake (25 per cent). Clearly what was left had to be divided between plastics (35 per cent) and processed oil (40 per cent). A rough approximation would make the processed oil share slightly bigger than the share for plastics.

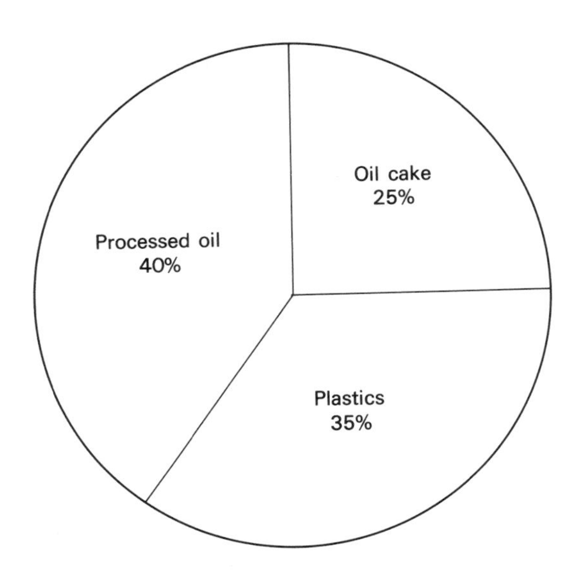

Pie diagram – company sales

Exercise 4.1.

A listing for three dice for 56 cases covers the following:

1,1,1; 1,1,2; 1,1,3; 1,1,4; 1,1,5; 1,1,6; 1,2,2; 1,2,3; 1,2,4; 1,2,5; 1,2,6; 1,3,2; 1,3,3; 1,3,4; 1,3,5; 1,3,6; 1,4,2; 1,4,3; 1,4,4; 1,4,5; 1,4,6; 1,5,2; 1,5,3; 1,5,4; 1,5,5; 1,5,6; 1,6,2; 1,6,3; 1,6,4; 1,6,5; 1,6,6; 2,2,2; 2,2,3; 2,2,4; 2,2,5; 2,2,6; 3,3,2; 3,3,3; 3,3,4; 3,3,5; 3,3,6; 4,4,2; 4,4,3; 4,4,4; 4,4,5; 4,4,6; 5,5,2; 5,5,3; 5,5,4; 5,5,5; 5,5,6; 6,6,2; 6,6,3; 6,6,4; 6,6,5; 6,6,6.

Exercise 4.2.

A listing of the cases of two dice covers the following:

For 21 (unordered) cases we have:

1,1; 1,2; 1,3; 1,4; 1,5; 1,6; 2,2; 2,3; 2,4; 2,5; 2,6; 3,3; 3,4; 3,5; 3,6; 4,4; 4,5; 4,6; 5,5; 5,6; 6,6.

For 36 (ordered) cases we have:

1,1; 1,2; 1,3; 1,4; 1,5; 1,6; 2,1; 2,2; 2,3; 2,4; 2,5; 2,6; 3,1; 3,2; 3,3; 3,4; 3,5; 3,6; 4,1; 4,2; 4,3; 4,4; 4,5; 4,6; 5,1; 5,2; 5,3; 5,4; 5,5; 5,6; 6,1; 6,2; 6,3; 6,4; 6,5; 6,6.

Exercise 4.3

The 216 ordered cases for three dice are:

1,1,1; 1,1,2; 1,1,3; 1,1,4; 1,1,5; 1,1,6; 1,2,1; 1,2,2; 1,2,3; 1,2,4; 1,2,5; 1,2,6; 1,3,1; 1,3,2; 1,3,3; 1,3,4; 1,3,5; 1,3,6; 1,4,1; 1,4,2; 1,4,3; 1,4,4; 1,4,5; 1,4,6; 1,5,1; 1,5,2; 1,5,3; 1,5,4; 1,5,5; 1,5,6; 1,6,1; 1,6,2; 1,6,3; 1,6,4; 1,6,5; 1,6,6; 2,1,1; 2,1,2; 2,1,3; 2,1,4; 2,1,5; 2,1,6; 2,2,1; 2,2,2; 2,2,3; 2,2,4; 2,2,5; 2,2,6;

2,3,1; 2,3,2; 2,3,3; 2,3,4; 2,3,5; 2,3,6; 2,4,1; 2,4,2; 2,4,3; 2,4,4; 2,4,5; 2,4,6;
2,5,1; 2,5,2; 2,5,3; 2,5,4; 2,5,5; 2,6,6; 2,6,1; 2,6,2; 2,6,3; 2,6,4; 2,6,5; 2,6,6;
3,1,1; 3,1,2; 3,1,3; 3,1,4; 3,1,5; 3,1,6; 3,2,1; 3,2,2; 3,2,3; 3,2,4; 3,2,5; 3,2,6;
3,3,1; 3,3,2; 3,3,3; 3,3,4; 3,3,5; 3,3,6; 3,4,1; 3,4,2; 3,4,3; 3,4,4; 3,4,5; 3,4,6;
3,5,1; 3,5,2; 3,5,3; 3,5,4; 3,5,5; 3,6,6; 3,6,1; 3,6,2; 3,6,3; 3,6,4; 3,6,5; 3,6,6;
4,1,1; 4,1,2; 4,1,3; 4,1,4; 4,1,5; 4,1,6; 4,2,1; 4,2,2; 4,2,3; 4,2,4; 4,2,5; 4,2,6;
4,3,1; 4,3,2; 4,3,3; 4,3,4; 4,3,5; 4,3,6; 4,4,1; 4,4,2; 4,4,3; 4,4,4; 4,4,5; 4,4,6;
4,5,1; 4,5,2; 4,5,3; 4,5,4; 4,5,5; 4,5,6; 4,6,1; 4,6,2; 4,6,3; 4,6,4; 4,6,5; 4,6,6;
5,1,1; 5,1,2; 5,1,3; 5,1,4; 5,1,5; 5,1,6; 5,2,1; 5,2,2; 5,2,3; 5,2,4; 5,2,5; 5,2,6;
5,3,1; 5,3,2; 5,3,3; 5,3,4; 5,3,5; 5,3,6; 5,4,1; 5,4,2; 5,4,3; 5,4,4; 5,4,5; 5,4,6;
5,5,1; 5,5,2; 5,5,3; 5,5,4; 5,5,5; 5,5,6; 5,6,1; 5,6,2; 5,6,3; 5,6,4; 5,6,5; 5,6,6;
6,1,1; 6,1,2; 6,1,3; 6,1,4; 6,1,5; 6,1,6; 6,2,1; 6,2,2; 6,2,3; 6,2,4; 6,2,5; 6,2,6;
6,3,1; 6,3,2; 6,3,3; 6,3,4; 6,3,5; 6,3,6; 6,4,1; 6,4,2; 6,4,3; 6,4,4; 6,4,5; 6,4,6;
6,5,1; 6,5,2; 6,5,3; 6,5,4; 6,5,5; 6,5,6; 6,6,1; 6,6,2; 6,6,3; 6,6,4; 6,6,5; 6,6,6.

Exercise 4.4.

The 25 cases which three dice produce 9 are:

2,2,5; 2,5,2; 5,2,2; 1,2,6; 1,6,2; 2,1,6; 2,6,1; 6,2,1; 6,1,2;
1,3,5; 1,5,3; 3,1,5; 3,5,1; 5,1,3; 5,3,1; 3,5,1; 2,3,4; 2,4,3;
3,2,4; 3,4,2; 4,2,3; 4,3,2; 3,3,3; 3,3,3; 3,3,3.

The 27 cases that produce 10 are:

2,3,5; 2,5,3; 3,2,5; 3,5,2; 5,2,3; 5,3,2; 2,4,4; 4,2,4;
4,4,2; 2,2,6; 2,6,2; 6,2,2; 3,3,4; 3,4,3; 4,3,3; 1,4,5;
1,5,4; 4,1,5; 4,5,1; 5,1,4; 5,4,1; 1,6,3; 1,3,6; 3,1,6;
3,6,1; 6,1,3; 6,3,1.

Exercise 7.1

The word count on page 95 shows that there are 348 actual words. These are set out in 36 lines. To find an approximation of the words on a page you can count the number of lines and multiply this by a rough average of the words per line (found by counting the words per line on three or four lines and selecting the most frequent number).

My calculation of the words on page 95 was found by multiplying the number of lines (36) by my selected 'average' words per line (10) to give a total of 360. The difference between this approximation and the actual number of words on the page (348) is 12. How close did you get?

Exercise 10.1

The square root of 8 = 2.828
The square root of 7 = 2.646
difference = 0.182

The square root of 8,000 = 89.442719
The square root of 7,999 = 89.437128
difference = 0.005591

By inspection, it is clear that 0.182 is a much larger difference than 0.005591.

Exercise 11.2

Starting with 25 marks whether you study or not (and assuming the relationship between studying and marks is given by:

$$Y = 25 + 2X$$

then you require to get another 25 marks to pass at 50. At two marks per hour of study you require 25/2 = 12.5 hours.

If you did 25 hours of study, you would get 25 x 2 = 50 marks and adding this to the 25 marks you get whatever you do, your total mark would be 75.

Index